BEGINNING STATISTICS
AN INTRODUCTION FOR SOCIAL SCIENTISTS

SAGE was founded in 1965 by Sara Miller McCune to support the dissemination of usable knowledge by publishing innovative and high-quality research and teaching content. Today, we publish more than 750 journals, including those of more than 300 learned societies, more than 800 new books per year, and a growing range of library products including archives, data, case studies, reports, conference highlights, and video. SAGE remains majority-owned by our founder, and on her passing will become owned by a charitable trust that secures our continued independence.

Los Angeles | London | Washington DC | New Delhi | Singapore

2ND EDITION

BEGINNING STATISTICS

AN INTRODUCTION FOR SOCIAL SCIENTISTS

LIAM FOSTER
IAN DIAMOND
JULIE JEFFERIES

Los Angeles | London | New Delhi
Singapore | Washington DC

Los Angeles | London | New Delhi
Singapore | Washington DC

SAGE Publications Ltd
1 Oliver's Yard
55 City Road
London EC1Y 1SP

SAGE Publications Inc.
2455 Teller Road
Thousand Oaks, California 91320

SAGE Publications India Pvt Ltd
B 1/I 1 Mohan Cooperative Industrial Area
Mathura Road
New Delhi 110 044

SAGE Publications Asia-Pacific Pte Ltd
3 Church Street
#10-04 Samsung Hub
Singapore 049483

Editor: Katie Metzler
Assistant editor: Lily Mehrbod
Production editor: Ian Antcliff
Copyeditor: Richard Leigh
Proofreader: Louise Harnby
Indexer: Silvia Benvenuto
Marketing manager: Ben Griffin-Sherwood
Cover design: Shaun Mercier
Typeset by: C&M Digitals (P) Ltd, Chennai, India
Printed and bound in Great Britain by
CPI Group (UK) Ltd, Croydon, CR0 4YYR

Library of Congress Control Number: 2014940764

British Library Cataloguing in Publication data

A catalogue record for this book is available from
the British Library

MIX
Paper from
responsible sources
FSC® C013604
www.fsc.org

ISBN 978-1-4462-8069-0
ISBN 978-1-4462-8070-6 (pbk)

At SAGE we take sustainability seriously. Most of our products are printed in the UK using FSC papers and boards.
When we print overseas we ensure sustainable papers are used as measured by the Egmont grading system.
We undertake an annual audit to monitor our sustainability.

To Andrea, Chloe, Kathy and Ian; Jane, Alexandra, Mark and Tom; and Brian and Jenny.

CONTENTS

LIST OF TABLES

LIST OF FIGURES

PERMISSIONS

The authors and publisher would also like to thank the following for permission to use copyright material:

Tables 2.1 and 2.2 from Lader, D. and Steel, M. (Office for National Statistics) (2010) *Opinions Survey Report No. 42. Drinking: Adults' Behaviour and Knowledge in 2009*. A report on research using the National Statistics Opinions Survey produced on behalf of the NHS Information Centre for Health and Social Care. London: Stationery Office. http://www.ons.gov.uk/ons/dcp19975_50795.xml. Adapted from data from public sector information licensed under the Open Government Licence v.2.0. Crown Copyright.

Tables 3.13 and 3.18, and Figures 3.19, 3.20, 3.21 and 3.22 from Lader, D. (Office for National Statistics) (2009) *Opinions Survey Report No. 41: Contraception and Sexual Heath, 2008/09*. A report on research using the National Statistics Opinions Survey produced on behalf of the NHS Information Centre for Health and Social Care. London: Stationery Office. http://www.ons.gov.uk/ons/rel/lifestyles/contraception-and-sexual-health/2008-09/2008-09.pdf. Adapted from data from public sector information licensed under the Open Government Licence v.2.0. Crown Copyright.

Tables 2.4, 2.5, 6.6, 6.7 and 6.8, and Figures 5.5 and 5.6 from Office for National Statistics (2013) *2011 Census: Key and Quick Statistics for Local Authorities in England and Wales*. Adapted from data from public sector information licensed under the Open Government Licence v.2.0. Crown Copyright.

Table 2.6 from Harker, R. (2012) *Children in Care in England: Statistics*, Standard Note SN/SG/4470. http://www.parliament.uk/briefing-papers/sn04470.pdf. Adapted from data from public sector information licensed under the Open Government Licence v.2.0. Crown Copyright.

Tables 2.7 and A2.1 from Office for National Statistics (2011) 'Population estimates by marital status', *Mid 2010 Statistical Bulletin*. http://www.ons.gov.uk/ons/dcp171778_244768.pdf. Adapted from data from public sector information licensed under the Open Government Licence v.2.0. Crown Copyright.

Tables 2.8 and A2.2 adapted from Worcester, R. and Herve, J. (Ipsos MORI) (2010) *Was it the Sun (and the Times) wot (Nearly) Won it?* http://www.ipsos-mori.com/newsevents/ca/506/Was-it-the-Sun-and-the-Times-wot-nearly-won-it.aspx. Reproduced with permission from Ipsos MORI.

Tables 3.1, 4.11 and A4.1, and Figures 3.1 and 3.3 from Park, A., Clery, E., Curtice, J., Phillips, M. and Utting, D. (2012) *British Social Attitudes: The 29th Report*. London: NatCen Social Research. http://www.bsa-29.natcen.ac.uk. Adapted from data from public sector information licensed under the Open Government Licence v.2.0. Crown Copyright.

Tables 14.11, 14.12, A14.1, A14.2 and A14.3 from Park, A., Bryson, C., Clery, E., Curtice, J. and Phillips, M. (2013) *British Social Attitudes: The 30th Report*. London: NatCen Social Research. http://www.bsa-30.natcen.ac.uk/. Adapted from data from public sector information licensed under the Open Government Licence v.2.0. Crown Copyright.

Table 3.2 and Figures 3.2, 3.4 and 3.5 adapted from Alzheimer's Society (2012) *Dementia: A National Challenge Report*. http://www.alzheimers.org.uk/site/scripts/download_info.php?fileID=1389. All figures, unless otherwise stated, are from YouGov Plc. Reproduced with permission from the Alzheimer's Society and YouGov Plc.

Tables 3.3, 3.4, 3.5, 3.6, 3.7, 3.15, 3.16, 4.9, 13.1, 13.5 and A3.1, and Figures 13.2, 13.10, A3.1, A3.2 and A13.2 adapted from and reproduced with the permission of the publisher from the World Health Organization (2012) *World Health Statistics 2012* (page 98 section 4, page 109 section 5, page 51 section 1, and page 158 section 9). Geneva: WHO. http://apps.who.int/iris/bitstream/10665/44844/1/9789241564441_eng.pdf?ua=1 (accessed 4 January 2014).

Tables 3.8 and 3.9, and Figures 3.6, 3.7, 3.8, 3.9 and 3.10 adapted from UNESCO Institute for Statistics (2013) http://www.uis.unesco.org/datacentre. Reproduced with permission from UNESCO.

Table 3.10 and Figures 3.10 3.11, 3.12 and 3.13 from Office for National Statistics (2013) *Families with Dependent Children by Number of Dependent Children in the Household, UK*, Produced by Demographic Analysis Unit, Labour Force Survey. Adapted from data from public sector information licensed under the Open Government Licence v.2.0. Crown Copyright.

Table 3.17 and Figure A3.3 adapted from Eurostat (2011) *Percentage of Women's Employment which is Part-time, in Four EU Countries, 2011*. http://epp.eurostat.ec.europa.eu/tgm/refreshTableAction.do?tab=table&plugin=1&pcode=tps00159&language=en. Reproduced with permission from the Publications Office of the European Union.

Table 3.18 and Figure A3.4 adapted from Eurostat (2013) *Acquisition of Citizenship* http://epp.eurostat.ec.europa.eu/tgm/table.do?tab=table&init=1&plugin=1&language=en&pcode=tps00024. Reproduced with permission from the Publications Office of the European Union.

Table 3.19 and Figure A3.5 from Department for Education (2012) 'Government publishes destination data for the first time', 17 July. https://www.gov.uk/government/news/government-publishes-destination-data-for-the-first-time. Adapted from data from public sector information licensed under the Open Government Licence v.2.0. Crown Copyright.

Table 4.3 and Figures 4.1 and 4.2 from Office for National Statistics (2013) *Crime Statistics, Focus on: Violent Crime and Sexual Offences, 2011/12*. http://www.ons.gov.uk/ons/publications/re-reference-tables.html?edition=tcm%3A77-290621. Adapted from data from public sector information licensed under the Open Government Licence v.2.0. Crown Copyright.

Table 4.7 adapted from Cribb, J., Joyce, R. and Phillip, D. (2012) *Living Standards, Poverty and Inequality in the UK: 2012*. Institute of Fiscal Studies Commentary C124. London: IFS. http://www.ifs.org.uk/comms/comm124.pdf. Reproduced with permission from the Institute of Fiscal Studies.

Table 4.8 adapted from Banks, J., Tetlow, G. and Wakefield, M. (2008) *Asset Ownership, Portfolios and Retirement Saving Arrangements: Past Trends and Prospects for the Future*. Consumer Research 74. London: Financial Services Authority. http://www.fsa.gov.uk/pubs/consumer-research/crpr74.pdf. Reproduced with permission from the Financial Services Authority.

Table 4.10 published with the permission of the Scottish Professional Football League, copyright and database rights owner.

Table 4.12 from Horsfield, G. (Office for National Statistics) (2011) *Family Spending. A report on the Living Costs and Food Survey 2010*. London: Stationery Office. http://www.ons.gov.uk/ons/rel/family-spending/family-spending/family-spending-2011-edition/index.html. Contains public sector information licensed under the Open Government Licence v2.0. Crown Copyright.

Tables 5.2, 5.3 and 9.1 from Trussell, J. and Westoff, C. (1980) 'Contraceptive practice and trends in coital frequency'. *Family Planning Perspectives*, 12(5): 246–249. Adapted and reproduced with permission from John Wiley & Sons Ltd.

Table 6.10 from the World Bank (2013) *World Development Indicators*. Washington, DC: World Bank. http://data.worldbank.org/. Reproduced with permission from the World Bank.

Table 9.2 from Vauclair, C.-M., Abrams, D. and Bratt, C. (DWP) (2010) *Measuring Attitudes to Age in Britain: Reliability and Validity of the Indicators*. DWP Working Paper No. 90. London: Department for Work and Pensions. https://www.gov.uk/government/uploads/system/uploads/attachment_data/file/214388/WP90.pdf

Adapted from data from public sector information licensed under the Open Government Licence v.2.0. Crown Copyright.

Tables 10.1 and 10.2 adapted from YouGov (2013) *Voting Intention*. http://cdn. yougov.com/cumulus_uploads/document/1or1j1cocr/YG-Archives-Pol-ST-results-03-050212.pdf and http://cdn.yougov.com/cumulus_uploads/document/ lxcfy3g2a1/YG-Archives-Pol-Sun-results-080212.pdf. Reproduced with permission from YouGov.

Table 12.4 adapted from Daycare Trust and Family and Parenting Institute (2013) *Childcare Costs Survey 2013*. London: Daycare Trust. http://www.daycaretrust.org. uk/data/files/Research/costs_surveys/Childcare_Costs_Survey_2013.pdf. Reproduced with permission from the Daycare Trust and Family and Parenting Institute

Table 13.1, and Figures 13.2 and 13.10 adapted from United Nations (2011). *2011 Update for the MDG Database: Contraceptive Prevalence*. Department of Economic and Social Affairs, Population Division. http://www.un.org/esa/population/ publications/2011-mdgdatabase/2011_Update_MDG_CP.xls. Reproduced with permission from the United Nations.

Table 13.3 and Figures 13.11 and 13.12 from Humby, P. (Office for National Statistics) (2013) *An Analysis of Under 18 Conceptions and their Links to Measures of Deprivation, England and Wales, 2008–10*. http://www.ons.gov.uk/ons/ dcp171766_299768.pdf. Adapted from data from public sector information licensed under the Open Government Licence v.2.0. Crown Copyright.

Tables 14.4 and 14.5 adapted from Ipsos MORI (2010) *2010 Election Aggregate Analysis*. Reproduced with permission from Ipsos MORI.

Tables A1.2 and A1.3 adapted from Fisher and Yates (1974); reprinted with the permission of Pearson Education Limited.

Every effort has been made to trace all copyright holders. If any have been overlooked, or if any additional information can be given, the publishers will be pleased to make the necessary arrangements at the first opportunity.

ACKNOWLEDGEMENTS

Ian and Julie would like to thank past members of the Department of Social Statistics, University of Southampton, for contributions they made, directly and indirectly, to the content of the first edition of this book. In particular, Steve Pearson, who read through some of the final first edition manuscript, and Dave Holmes for his useful advice.

Liam would like to thank members of the Department of Sociological Studies, University of Sheffield, for thoughts and ideas in relation to the content of the second edition of this book. In particular, the useful insights of Tom Clark and Mark Tomlinson.

A big thank you also goes to the students we have taught, for their many comments, criticisms and suggestions about how to improve the teaching of quantitative methods.

Finally, thanks to Katie Metzler and Lily Mehrbod at Sage for their patience, feedback and guidance throughout the process of putting together this second edition.

ONE

INTRODUCTION – ARE STATISTICS RELEVANT TO REAL LIFE?

Introduction

The first edition of *Beginning Statistics* by Ian Diamond and Julie Jefferies stemmed from a belief that there were some excellent general statistics texts available at the time but none of them had the right tone or coverage of social statistics. Widely used across social science courses in particular, the first edition received much positive feedback. Despite its success, it is evident that it was becoming dated given that it is now over 10 years old and, as such, a number of the examples and features required updating. The decision to write the second edition, with the assistance of Liam Foster, also comes at a time when we are witnessing an increasing emphasis on the need for social science students to possess statistical research skills. This new edition provides an opportunity to bring data sources up to date, to clarify and add some explanations, provide chapter learning objectives and conclusions and a glossary. It also contains an expanded introduction, new sections on sampling and presentation conventions, and a conclusion. Despite these changes we still start with the basic assumption that you may have done no or limited previous work with statistics or may require a refresher. Maths may also be a distant (and painful) memory. By making use of interesting examples (well, we think so!) from the social sciences this book introduces a number of statistical approaches and techniques which will be invaluable to the development of your statistical skills. It does this in a systematic way, taking you through the steps required to use statistics in the social sciences, from introducing data and their presentation, to correlation and regression.

Did you know you already use statistics?

Believe it or not, you can *already* think **statistically** – in fact, it is something you do all the time. If you don't believe this, think about the number of times you have thought about the average number of hours you are studying or working in a week, how many times you have decided to take a coat with you because it is likely to be cold this time of year, or noted the fact that you tend to do better when you start revising for your exams earlier! In each case, you have acted on the basis of statistical concepts and used data that you had stored in your brain. You just didn't realise it! You are making a statistical statement even though you are performing at best rough calculations. In the first example you are summarising data; in the second and third you are generalising from previous experiences of weather patterns and exam performance to make a prediction or statement.

So our observations often involve, for instance, the process of counting how many times something has occurred, and measuring the length of time since something has happened. As if, by instinct, we look for patterns and connections among the things we notice, often in a rather imprecise manner. This might be something as simple as whether you are more likely to score a goal when wearing lucky pants or have a more successful date when you go for a meal, bowling or ice-skating! We constantly question data and use them to influence our decisions and behaviour. This is where statistics are crucial. They are a way of making sense of our observations.

Learning about statistics helps you to look for reliable patterns and associations in both the short and long term (Rowntree, 2003). It also teaches us caution in expecting these to hold true in all situations (for instance, unfortunately Liam doesn't always score a goal with his lucky pants on). There are often important limitations to data which need to be considered, including whether they are biased, unrepresentative or totally meaningless (it does happen!). It teaches us to think critically about our techniques, samples and claims we make. Statistics is basically about understanding and knowing how to use **data**. So we all use statistical concepts intuitively in our daily lives. Learning statistics is simply a case of learning how to express things more accurately. By using statistical concepts this enables us to summarise and predict more precisely than we normally would in our everyday observations.

How are statistics used?

Data provide information which governments and organisations use to make policy decisions (and to evaluate the effectiveness of existing policies). Statistics about institutions such as schools and hospitals are increasingly collected and made available to the general public. This may be useful in deciding where a child is going to have the best chance of getting their A levels, or where you are least likely to die in the operating theatre. Such league tables can increase transparency and choice, but they may also be seen to fulfil a political purpose.

This century has been called the century of 'big data', and the techniques in this book are ever more important in this context. 'Big data' refers to every piece of knowledge that has or will be digitalised and stored on a computer hard drive, a database, or in the 'cloud'. It is vital to millions of companies such as Facebook, Google and Twitter, as well as Tesco, Nectar and Amazon, which harvest user information, making use of big data in targeted sales and advertising. For instance, if you have been looking for a nice pair of walking boots you are more likely to receive adverts about tents or camping holidays because statistically those who buy walking boots are more likely to go camping than those who don't. Simple stuff, I know, in an example such as this, but statistics are being used to look at these kinds of relationships in buying behaviour. So it is not only governments that use data to monitor and analyse behaviour.

There are two main kinds of statistics that we are going to concentrate on in this book. **Descriptive statistics** is a set of methods used to describe data and their characteristics. For example, if you were investigating the number of visitors to a beach in August (nice job if you can get it!), you might draw a graph to see how the number of visitors varied each day, work out how many people visit on an average day and calculate the proportion of visitors who were male/female or children/adults. These would all be descriptive data.

Inferential statistics involves using what we know to make inferences (estimates or predictions) about what we don't know. For example, if we asked 200 people who they were going to vote for on the day before a local election we could try to predict which party would win the election. Or if we asked 50 injecting drug users whether they share injecting equipment such as needles with other users, we could try to estimate the proportion of all injecting drug users who share equipment.

We would never be able to say for sure who would definitely win the election or what proportion of injecting drug users share equipment, but we *are* able to predict the *likely* outcome or proportion. Statistics is all about weighing up the chances of something happening or being true. Statistics *are* relevant to real life because without real life we wouldn't need statistics. If everybody always voted for the same party and all injecting drug users shared equipment, we wouldn't need to predict the outcome of an election or estimate the proportion of drug users sharing equipment because it would be obvious from asking one voter or one injecting drug user. Only in a world of clones would statistics about people be unnecessary! Life would be pretty boring if everybody were exactly the same.

So in real life everybody is different and, in social science, statistics are frequently used to highlight the differences between groups of people or places. For example, we might want to investigate how smoking behaviour varies by socio-economic group or how unemployment varies by local authority. Knowledge of statistical methods is crucial for answering many research questions like these.

The emergence of statistics

We know that statistics are commonly used in contemporary society, but where did they come from? The use of statistics is nothing new. It goes back to at least the earliest city states. For instance, Babylonians and Egyptians collected numerical data on crops and growing conditions. The word 'statistics' is derived from the Latin term for 'state' or 'government'. In the UK and the Westernised world it was particularly the birth of industrialism that led to an interest in social data. In Europe and the USA censuses started to be taken in the eighteenth century, with the first in the UK in 1801. In the UK a census has been conducted every 10 years (with the exception of 1941 as a result of the Second World War).

Sapsford (2007), commenting on the history of statistics in the UK, stated that, for early Victorians, the role of statisticians was to collect information about people in the emerging capitalist societies. It was thought that we needed to 'map' the human population to make the best use of people in industry as well as providing the services they needed. It was important to know about things like where people lived, their ages (including how many were able to work), what children were living in the family, whether there were many older or disabled people who needed support, and what types of housing were available.

The potential of these kinds of data for social scientists was shown by people such as Charles Booth (1886), whose work on occupation patterns was derived from analysis of the 1801–1881 UK censuses. The first journal of the Royal Statistical Society in 1838 had lots of articles describing the social conditions of the time. Large-scale social surveys looking at urban poverty were pioneered by Seebohm Rowntree (1901) in York and Charles Booth (1902) in London at the turn of the twentieth century, as statistics started to play a greater role in exploring social problems. Following the Second World War, statistical methods have enjoyed increasing popularity in the social sciences. There has also been an increase in the number of large scale national datasets available, covering a wealth of areas such as health, crime, employment, housing and social attitudes. These have been used by academics and policy-makers alike.

This increase in the number of datasets available, both national and international, longitudinal and cross-sectional, has been accompanied by advances in technology, in particular the influx of computers, especially from the 1980s and 1990s. With this came further access to data analysis for a larger number of people and further possibilities in its use. Specific computer programmes such as IBM SPSS, SAS, STATA and R have been developed which make the analysis of statistical data more accessible to social scientists. Analysis is now possible at the click of a button (although you need to know which buttons to press and what the results mean!). Statistical tests have continued to be developed enabling higher levels of analysis. This has been supported by the UK Data Service (formerly the UK Data Archive), which helps researchers to acquire data and supports their use. Despite this, unfortunately it is not uncommon for statistics courses taken by students in the social sciences to be treated essentially as maths courses with examples used which do not relate to the social sciences. This can be off-putting to those learning statistics and something this book is mindful of.

Do we really need to know about statistics?

Basically, if data are to be useful, they have to be processed and analysed. This requires statistical skills, which will become ever more important in the social sciences and beyond. Many of the concepts that underlie statistical analysis are not particularly complex and are things which people do on a daily basis, as we have already shown. The majority of people have some understanding of figures (even if they don't admit it). They know a 10% pay rise is better than a 5% one or can work out the odds on the horse they bet on winning. This will stand you in good stead for this book and for developing your statistical capabilities.

There are two major reasons why learning about statistics will be useful to you:

- You are constantly exposed to statistics every day of your life, as you have already seen. Marketing surveys, voting polls and findings from social research appear in daily newspapers and popular magazines. By learning about statistics you will (hopefully) become a more effective consumer of statistical information. For example, if a hair advert tells you that 75% of participants say their hair is silky and smooth following the use of a particular shampoo you would start to question what they were using before (or not using before!) and think about the sample size (if the small print says a sample of 10 you might be a bit worried!).
- You need to be able to understand and interpret statistics at university or in the workplace. Even if conducting research is not part of your job or you don't do a quantitative project at university, you will be expected to understand and learn from other people's research based on statistical analysis.

Modern society is driven by statistics which frequently influence our behaviour, from reviews on TripAdvisor which tell us when a place really is a dive which we need to avoid like the plague, to crime statistics which tell us which areas we are most likely to be burgled in (and might not want to move to if we have a choice in the matter!). Even if you never go on to do research, a good grasp of statistics will help you to understand the figures that you read or hear about and to avoid being misled by people who (mis)use statistics to their own advantage.

The need to promote statistical skills among social scientists is perhaps greater than ever. Recent discussion has recognised that many graduates in the social sciences lack the quantitative skills required by the social research industry and that quantitative skills are in demand by employers (Wiles et al., 2009). This has led to lots of institutions working to improve their quantitative research teaching and research. This has also been encouraged by Q-Step, a £19.5 million programme funded by the Nuffield Foundation, ESRC and HEFC to set up 15 quantitative research centres at UK universities designed to promote a step-change in quantitative social science training.

'There are three kinds of lies: lies, damned lies and statistics'

The expansion of data available has also led to increasing debate about how figures are constructed. While numbers are often thought of as hard facts, they are actually the result of different decisions about how something should be categorised or counted. There is much cynicism about statistics and how they are used. Mark Twain reported that the British statesman Benjamin Disraeli once said, 'there are three kinds of lies: lies, damned lies and statistics'. More recently the postmodern theorist Jean Baudrillard (1990: 147) said, 'like dreams, statistics are a form of wish fulfilment'. However, to reject statistics would involve blinding oneself to much important information.

It is crucial to be aware that statistical data can be misinterpreted, distorted or selected to serve particular ends. This is not the inherent fault of statistics per se; rather the fault of analysis which does not carefully examine the logic of an argument and how data support this (Dietz and Kalof, 2009). Not all inaccurate use of statistics is deliberate (not everybody has sinister motives – I can assure you!). Things can go wrong when figures are written down inaccurately, files not backed up, research poorly designed and inappropriate statistical tests used. Sometimes it is how the research is reported that is problematic. For instance, it has been known for the media to report statistics in a way that comes across as misleading or politicians to be very selective about the figures they present.

It is also common for inaccurate conclusions about findings to be made (and bits that don't fit left out!) without looking at whether it actually means what is stated or whether there could be other possible explanations! So, for instance, several years ago research by Halpern and Coren (1991) found that the average age of death of left-handers was about 8 years less than for right-handers. Data were collected by using a sample of 987 deaths in Southern California (where lists are published of everyone who has died) and interviewing relatives or friends to find out the hand they used most often. As a left-handed person, Liam was particularly interested (and worried) about this research and had a closer look at it. The authors of the study speculated why left-handed people may be more likely to die at a younger age, but Strang (1991) stated that they failed to sufficiently account for possible bias given that the frequency of left-handedness is likely to have changed appreciably over time. For instance, if many people born 60 or more years ago (such as Liam's mum) were forced to use their right hands, then this would result in a large number of right-handed people being reported as dying in older age, with the probability of being left-handed greater among younger people, resulting in a lower average age of death for left-handed people. Whether this explains the eight year difference is another question! Perhaps a bigger issue is that the study was based on a sample of deaths and didn't actually measure the chance of left or right handers in the population dying at different ages.

Another example commonly referred to is the idea that children who eat breakfast (not a chocolate bar from the local corner shop though) perform better at school in general than those who don't eat their bowl of Weetabix or turn the milk chocolatey with their Coco Pops. While it is true that statistics do indicate a correlation between having breakfast and academic achievement (leading to parents making their children hear that snap, crackle and pop in the morning), when the trend was investigated, although food can aid concentration, it was found that it was largely the types of children who have things going on in their lives which stop them from eating breakfast who are also more likely to struggle at school.

A second problematic example involving children (Levitas, 2012) is the coalition government's 'Troubled Families' initiative, targeted at those families who 'fail to take responsibility for their own lives'. Current government policy on social justice claims that there are 120,000 'troubled families' in Britain. The Department for Communities and Local Government (DCLG, 2013) identifies the research the figure is based upon (though not the details of the costing). The original research is a report carried out for the Social Exclusion Task Force in 2007 using secondary analysis of the Family and Children Study, a longitudinal survey. It focused on families with troubles rather than 'troubled families'. This analysis showed that in 2004 about 2% of the families in the survey had five or more of seven characteristics, and were severely multiply disadvantaged. The characteristics were: no parent in the family is in work; the family lives in overcrowded housing; no parent has any qualifications; the mother has mental health problems; at least one parent has a long-standing limiting illness, disability or infirmity; the family has low income (below 60% of median income); and the family cannot afford a number of food and clothing items. That 2% of families generated an estimate of 140,000 for Britain, later calculated as 117,000 for England, rounded to 120,000. The DCLG website makes a jump from families that have troubles, through families that are 'troubled', to families that are or cause trouble (Levitas, 2012). Portes (2012) points out that none of these criteria, in themselves, have anything at all to do with disruption, irresponsibility, or crime. While a family meeting five criteria is likely to be disadvantaged and poor, are the criteria identified a source of wider social problems? At another point the DCLG (2012) stated that these families are characterised by being involved in crime and anti-social behaviour; have children not in school; have an adult on out-of-work benefits and cause high costs to the public purse, but still used the same figure of 120,000 despite it being based on data from 2004 and changing in definition! It is highly unlikely that the figures remain exactly the same (see Levitas, 2012). Therefore, these figures should be used with much caution (if at all)!

A final example that we like comes from Huck and Sandler (1979) and relates to a billboard advertising campaign and Miss America. In an attempt to prove that billboard advertising is the best form of advertising, the Institute of Outdoor Advertising conducted a research study. Using 10,000 billboard panels, they placed a poster showing a large picture of Miss America with her crown and the simple message, 'Shirley Cothran, Miss America, 1975'. Before they did this a series of studies were conducted to determine public awareness of Miss America's name prior to it going on the billboards, with a random sample of over 15,000 adults questioned in 1975. Despite previous exposure which Miss America had received on TV and radio and in print, only 1.6% of respondents gave the correct answer when asked, 'what is the name of Miss America 1975?' The billboards then went up and two months later a second wave of over 15,000 interviews was conducted by the same research teams (with different participants). This time, 16.3% of the respondents knew who Miss America was. This was said to indicate that the billboard advertising had a very positive effect on promoting the name of Miss America and was therefore a highly effective form of advertising. However, could there be any other reasons for this trend? Well, first, the study fails to acknowledge that other forms of media also existed during the time the billboards were up. Second, the unique nature of the billboard study led other forms of media to cover the story of the research, with Miss America's name discussed more than usual in the media. So even the results from a simple statistical question like the one asked here can be difficult to interpret and require some critical thinking, something this book is going to help you with. So an awareness of statistics helps to identify these types of concerns and mistakes, and many more like them!

Overview of the chapters

Statistics is rather like building a house (but not as physically demanding). It involves laying the foundations first, establishing the basics of using and presenting statistics, before more advanced statistics can be used (or the rest of the house can be built). Without this secure foundation and knowledge of descriptive statistics it is difficult to know what inferential statistics are appropriate, and this is likely to result in inaccurate statistics to interpret. Similarly, if you were building a house and the basic foundations were not established you risk the house falling down (and getting sued if it caused other damage!). This is where the book starts, gradually moving from simple topics to the more complicated ones.

Chapters 2 and 3 focus on how data are measured, or how the levels of measurement affect how they can and should be presented, with a particular focus on different forms of tables and graphs. The correct type of table or graph to use depends on the type of data you have, not just because it looks pretty! There are a number of rules or conventions you need to be familiar with in the design and presentation of tables and graphs which are covered in these chapters.

Chapters 4 and 5 move on to look at ways of describing data. Rather than concentrating on investigating the distribution of data by drawing graphs and tables, these chapters explore how to describe a dataset statistically, so that we can summarise the features of a distribution (descriptive statistics). They focus on measures of central tendency including the mean, median and mode, and calculating and presenting percentiles, terms you will become familiar with. These are useful in helping us to identify key patterns such as the most common response to a question or the average number of times an event occurs over a period of time.

Sometimes in order to use data more effectively it is necessary to transform them. Chapter 6 explores how and when this can be useful. It may be something as simple as standardising measurements when you are provided with information in both miles and kilometres and you want to undertake research on distance travelled to work. Transforming variables can be a useful thing to do when dealing with statistics. This includes the process of standardising the data and producing Z-scores.

Chapter 7 introduces the process of calculating whether data are normally distributed and, building on the previous chapter, how measures of standard deviation and Z-scores are used to do this. Inferential statistics are used as we show how we can predict the probability of an event occurring when data are normally distributed. In Chapter 8 the focus turns to sampling and how to select a representative sample in order to make estimates of the things we are trying to find out about in the population. Using a sample can save vast amounts of time and money. This chapter highlights different forms of sampling and when they might be used. It shows how it is not always as easy as it sounds and takes you through strategies to help with representativeness.

Chapters 9 and 10 continue to focus on the use of samples and their relationship with populations, using confidence intervals. However, Chapter 9 uses continuous data whereas Chapter 10 deals with proportions.

Chapters 11 and 12 provide a brief introduction to hypothesis testing in various contexts. This involves using confidence intervals and Z-scores, when you have sample means or proportions, to test hypotheses (in Chapter 11). It also shows the difference between two-sided (or two-tailed) tests and one-sided (or one-tailed) tests in calculating whether a hypothesis should be proved or disproved. Chapter 12 introduces the t test, used to assess the statistical significance of the difference between the means of two sets of scores and whether the average score for one set of scores differs significantly from the average score for another set of scores. It also provides details about when Z-scores, one-sample t tests and one-sample sign tests in particular should be used.

The process of comparing two variables with each other is the main focus of Chapter 13. For instance, is a student's performance at university related to how much paid work they are doing? To assist in answering this kind of question, correlation can be used to measure the association between two continuous variables. Regression takes things a little further, enabling us to predict the values of one variable from the values of another variable. For instance, the data on the number of qualified doctors in countries per 1000 people could be used to try to predict the death rate. It is also possible to use inferential statistics to analyse categorical data such as using the chi-square test, which allows us to test the association between two variables in which the expected values are compared with the observed values. These processes are shown in Chapter 14. Chapter 15 provides you with more information about how and where to develop your statistical skills further, something we hope you will want to do when you have completed this book.

TWO
DATA AND TABLE MANNERS

Introduction

Before starting to do exciting things with data, you need to know a bit more about them. For instance, how data are measured, or how the levels of measurement affect how they are presented and what you can do with them. It is also important to understand how to present data. You could have some really interesting findings, but if they are not presented clearly they are of little or no use. Also, the way in which data are presented and reported can affect how they are interpreted. For instance, something as simple as whether numbers or percentages are used can create a very different picture. This chapter will introduce you to using data and their presentation in the form of tables. It will highlight some areas of good practice in presentation of data which will be useful when you produce your own tables! By the end of the chapter you should be able to:

- Identify different levels of measurement and associated terminology
- Construct tables in a clear and appropriate manner
- Work out and use percentages
- Understand some of the issues with reporting data

Data

Let's look at what data are, as you need to know a few bits of jargon so that you know what to call things. This will also help when it comes to understanding and analysing tables. Suppose you have done a survey of students in a university students' union on a Friday afternoon (our experience is that this is the best place to find students on a Friday!), asking them which area of the town they live in and how much rent they have to pay each week. The two topics that you are interested in finding out about, 'area' and 'rent', are called the **variables**. Each variable has **attributes**. So if we take the variable 'do you believe in sex before marriage?', the attributes of the variable may be 'yes' and 'no'. Each student questioned is called a **case** and the responses that you get from each student are known as **observations** (even if you don't physically 'observe' them yourself). Data, then, are simply a collection of observations.

Whatever a set of data (a dataset) may refer to, the cases are the individuals in the **sample** (these may be people, countries, cars and so on), while variables are the characteristics which make the cases different from each other (for example, age, opinion about a topic, type of political system).

There are an infinite number of possible variables that we might be interested in and, in fact, variables themselves can be divided into several types.

Continuous variables include things like 'weights of newborn babies', 'distance travelled to work by those in full-time employment', or 'percentage of children living in lone-parent families'. Such variables are measured in numbers, and an observation may take any value on a continuous scale. For example, distance travelled to work could take a value of 0 miles for people working at home, 1.6 miles, 4.8 miles, or any other value up to 100 miles or more for those commuting long distances. Similarly, any variable measured as a percentage can take a value of 0%, 100% or anything in between. For continuous variables the standard rules of arithmetic apply, so it makes sense to say that if you commute for 4 miles then that is twice the commute of someone who commutes 2 miles.

Discrete variables or **categorical variables** are not measured on a continuous numerical scale. Examples of discrete variables are:

- sex: female/male
- religion: Buddhist/Christian/Hindu/Jewish/Muslim/Sikh/other
- degree subject studied: Politics/Sociology/Social Work

and so on. Such variables have no numeric value. We may assign them a number, for example, Politics = 1, Sociology = 2, Social Work = 3 and so on, but the actual numbers do not mean anything. For example, Social Work is not three times greater than Politics!

Most variables can be easily classified into continuous or discrete, although there are a few grey areas.

Whenever you collect or use some data, you should write down a clear **code book** to describe your variables. A code book gives information about each variable,

Variable name	Variable type	Units of measurement/codes
Sex	Categorical	1 = male 2 = female
Weight	Continuous	Kilograms
Attendance	Continuous	Percentage of lectures attended in semester
Political affiliation	Categorical	1 = Conservative 2 = Labour 3 = Liberal Democrat 4 = other 5 = none

such as name and type, along with the units of measurement or categories, as in the previous example.

It is also possible to split categorical variables further into different levels of measurement: nominal, dichotomous or ordinal.

- **Nominal** variables are variables that have two or more categories, but which do not have an intrinsic order or inherent numerical quality in themselves. So nominal variables include things like 'marital status', 'ethnicity' or 'location', in addition to the variables already identified including 'religion' and 'degree studied'. In some cases there may be many attributes to a nominal variable. For instance, if we were classifying where people live in the USA by state there would be 50 attributes (or states).
- **Dichotomous** variables are nominal variables which have only two categories or levels. For example, if we were looking at gender, we would generally categorise somebody as 'male' or 'female'. This is also a nominal variable. A further example would be if we asked somebody if they had ever smoked, giving them the possible answers of 'yes' or 'no'.
- **Ordinal** variables have two or more categories, like nominal variables, but the categories can also be ordered or ranked moving from greater to smaller values (or vice versa). So an opinion poll might ask how likely you were to vote for a particular party at the next election with the possible options 'very likely', 'likely', 'not sure', 'unlikely' or 'very unlikely'. While the responses are in a particular order we cannot (or should not!) place a 'value' on them. This is despite our own views about which we prefer! Another example would be if you asked someone to provide an answer to the statement 'I generally eat healthily' and had a similar scale from 'strongly agree' to 'strongly disagree'. It is difficult to say whether the distance between the two categories is equal as it will depend upon individual perception. So someone who has their 'five-a-day' and five chocolate bars too may consider themselves to eat healthily, whereas another person may consider that eating two chocolate bars in addition to lots of fruit and vegetables means they do not eat healthily. As a result, the distance between the points on the scale is not clear and continuous.

Continuous variables can be further categorised as either interval or ratio variables. These two terms are rarely used in social statistics, but for completeness in this book we will describe them briefly here.

- **Interval** variables can be measured along a continuum and have a numerical value. The distance between the ranks/attributes is the same but it has an arbitrary zero point. A highly familiar example of an interval scale measurement is temperature on the *Celsius scale*. Although it is based on the freezing point of water, the zero point could just as easily be based on the freezing point of alcohol. In any case, interval variables are rarely used in social research.
- **Ratio** variables have all the properties of an interval variable, but also have a clear definition of zero. Age, for instance, has a logical zero point, and can often be considered to be a ratio variable. Other examples of ratio variables include height, weight, distance and temperature when the Kelvin scale is used (this has an absolute zero point). The name 'ratio' reflects the fact that you can use the ratio of measurements. So, for example, a weight of 20 kg is twice the weight of 10 kg. Most interval level variables within social statistics are actually also ratio variables. In fact, we can't actually think of one that you are likely to come across in social research that isn't, but many textbooks will still mention it.

Tables

Now we know a bit more about types of data, let's think about how they should be presented. Tables are one of the best ways of presenting a set of data. You may be thinking, 'anyone can draw a table – don't insult my intelligence!' It is true that anyone can throw some figures and a few lines on to a page, but to produce a good table requires a little more thought. The aim of drawing a table is to transform a set of numbers into a format which is easy to understand.

Table 2.1 Alcohol consumption

Age	M	F
1	17.5	11.00
2	15	10.2
3	16.80	10.5
4	12.50	10.5

Look at Table 2.1, which uses findings by Lader and Steel (2010) based on figures from the Office for National Statistics Opinions Survey in Great Britain. What is wrong with this table? A few of the problems are listed as follows, and you may be able to find more:

- The title is uninformative. Who do the data refer to?
- The variables are not defined, so we do not know, for example, what age '1' refers to. Does it refer to the alcohol consumption of 1-year-olds?!
- We do not know what units the data are measured in: is it number of drinks per week or number of gallons per month or something completely different?
- The layout is not very helpful. For example, the columns of data are not aligned with the column headings, there are no lines (rules) in the table, and the decimal points are not lined up.
- The number of decimal places is not the same throughout the table.
- We are not told where the sample came from or how many people were included.

An improved version of the same table might look something like Table 2.2. We have used annotations to help you see which characteristics are needed to ensure the table layout is logical and easily grasped by the reader. As you can see, this is far more informative and the patterns shown in the data are immediately clear. You can see that on average males drink more units each week than females of comparable ages and that the 16–24 age group drink the most units of alcohol in Britain. These are interesting findings and ones you could explore further either through empirical research or in the literature.

You can see that in addition to the mean weekly alcohol consumption, the number of people responding in each group and the total number of males and females in the sample are listed. While this is not compulsory, it is often useful to make people aware of the sample size as this can affect the potential reliability/ validity of the research. If it was based on a small sample you would need to think about the extent to which it would be appropriate to generalise to a wider population. For instance, if you surveyed 20 prisoners about their experience of treatment in Holloway prison you could not simply claim that these findings are representative of the whole prisoner population. You need to be rather careful about the assumptions you make about your research findings anyway, but particularly cautious if you are dealing with a small sample size.

Table 2.2   Average weekly alcohol consumption in units[1] by sex and age, Great Britain, 2009

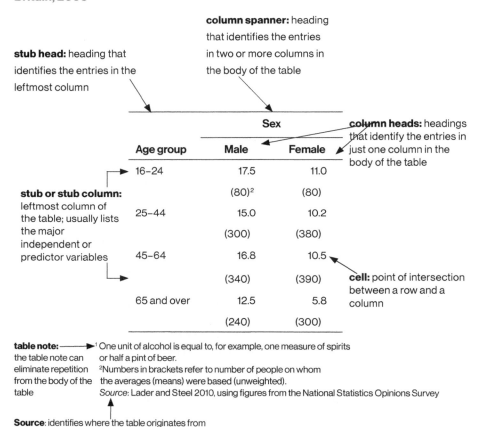

Age group	Sex	
	Male	Female
16–24	17.5	11.0
	(80)[2]	(80)
25–44	15.0	10.2
	(300)	(380)
45–64	16.8	10.5
	(340)	(390)
65 and over	12.5	5.8
	(240)	(300)

column spanner: heading that identifies the entries in two or more columns in the body of the table

stub head: heading that identifies the entries in the leftmost column

column heads: headings that identify the entries in just one column in the body of the table

stub or stub column: leftmost column of the table; usually lists the major independent or predictor variables

cell: point of intersection between a row and a column

table note: the table note can eliminate repetition from the body of the table

[1] One unit of alcohol is equal to, for example, one measure of spirits or half a pint of beer.
[2] Numbers in brackets refer to number of people on whom the averages (means) were based (unweighted).
Source: Lader and Steel 2010, using figures from the National Statistics Opinions Survey

Source: identifies where the table originates from

If the same table was produced using figures from a sample of only 18 people (Table 2.3) you can see how the impact of findings which may not be typical can affect the mean. So for instance, the two 45–64-year-old males in these hypothetical results had not consumed alcohol in the last week; perhaps they didn't drink at all anyway or were on antibiotics at the time of the survey. This compares with a mean of 16.8 among this group based on a sample of 340 in Table 2.2. You really would need to question the findings from Table 2.3 and ask yourself whether it tells you much other than what 18 people have said about their drinking behaviour. Of course, what actually constitutes a large sample is relative to the population. You will learn more about sampling later.

Table 2.3 Average weekly alcohol
consumption in units[1] by sex and age

Age group	Sex	
	Male	Female
16–24	25.0	5.0
	(2)[2]	(1)
25–44	15.0	10.5
	(2)	(2)
45–64	0.0	20.0
	(2)	(4)
65 and over	12.0	13.0
	(2)	(3)

[1]One unit of alcohol is equal to, for example, one measure of
spirits or half a pint of beer.
[2]Numbers in brackets refer to number of people on whom the
means were based (unweighted).

Source: Hypothetical data

Table rules

Tables can help you present a large amount of material efficiently. However, table layout needs to be logical and easy for readers to understand. There are a number of key points to remember when drawing tables.

- Always have a clear title: who/what, when and where do the data refer to? You will usually need a numbering system too, such as Table 1, Table 2 and so on.
- Make sure that all columns and rows are named properly, for example 'Male' and 'Female', not 'M' and 'F' (although you may understand it, your readers may not).
- Remember to state the units of measurement used, for example '%', '£000s'.
- If further explanation is needed to clarify certain points, put notes below the table, as in Table 2.2. These can be really useful for getting rid of repetition from the main body of a table.
- Include the source of the data: did you find them in a book or journal, or did you collect them yourself?
- Take care with layout and presentation: the table should be easy to understand. The use of lines within a table is a matter of preference, but do not put lines around every single value or the table may just look cluttered!
- For the same type of data, use the same number of decimal points throughout. One or two decimal places should be enough: if you use more you must have a good reason! Line up decimal points vertically. In this book, there are also a number of occasions where the information has been reproduced from elsewhere and the information was not available in decimal places.
- Include relevant totals and subtotals in the table and always check that they add up correctly. If you have been rounding up to two decimal places, for example, the totals may be slightly out, but do not worry too much about this. It may be an idea to indicate in a note below the table that this is the case.
- If you have a lot of data, it may be difficult to include all the data without the table becoming very complicated and difficult to read. You could try joining some categories together (known as collapsing categories), but bear in mind that you will lose some precision in doing this. You should not collapse categories just because there are small numbers in the cells of the tables unless it would breach the confidentiality of the respondents not to do this.

In general, tables must stand alone. This means that somebody should be able to understand a table without having to read the text before or after it. However, you should always refer to a table in the writing.

American Psychological Association conventions

Many publishers and courses specify that authors should conform to the American Psychological Association (APA) conventions, which were first introduced in 1929, when it comes to producing tables and figures. The APA produces a *Publication Manual*, which is now in its sixth edition (2010), and also has its own website (http://www.apastyle.org/). This provides (very detailed!) guidance on tables and figures, which builds on much of what we have said already. Some handy rules and tips from the APA (2010) are listed below:

- Use fonts that are large enough to be read without the use of magnification. Nobody likes to strain to see small figures! (In general, the APA recommends the use of 12-point type and double-spacing, although this may be adjusted for clarity, for example, to keep the table on one page.)
- Include all of the information needed to understand the table within it. This means that you should avoid novel abbreviations (which you might understand but readers don't), use table notes, and use clear labels.
- Keep tables free of unnecessary materials, no matter how decorative those materials may make the graphic look. It is not a piece of art but a way of presenting information in a clear and concise manner.
- Use Arabic numerals in the order in which they are first mentioned in the text. Never use suffix letters such as 5a and 5b to number tables and figures. If the work has an appendix with tables or figures, identify these parts of the appendix with capital letters and Arabic numerals. For instance, Table A1 would be the first table of Appendix A.
- Only use a capital letter for the first letter of the first word of all headings (unless it includes proper nouns – names, etc!).
- Be selective in terms of deciding the appropriate number of tables (and other figures) to include in your work. A large number of tables may result in the message getting lost. Think carefully about whether tables are always the most effective form of communication.
- Think carefully about where tables go in your piece of work, checking any specific guidelines.

Percentage tables

Table 2.2 contained a set of means (mean weekly alcohol consumption). Many tables simply consist of the number of people in each category. If this is the case, it may be useful to turn the numbers into percentages, so that two or more groups with a different total number of people can be compared. This is fairly straightforward, but the following example, Table 2.4, illustrates a hidden pitfall. If you have forgotten or never been taught how to work out percentages, don't worry, help is at hand below. If you remember it well then do skip the following description.

Proportions and **percentages** are very similar, the main difference being that a proportion can take any value from 0 to 1 inclusive, while a percentage usually takes a value from 0% to 100% (although percentages greater than 100% are possible).

To change a proportion into a percentage or vice versa, simply follow the rules below:

Proportion → percentage: multiply by 100
Percentage → proportion: divide by 100

For example:

- To turn the proportion 0.56 into a percentage, multiply by 100: 0.56 × 100 = 56%.
- To turn the percentage 48.5% into a proportion, divide by 100: 48.5 ÷ 100 = 0.485.

Notice that when you do this the numbers stay the same and just the decimal point moves.

Usually, life is not this simple and you will have to start from scratch with some numbers. For example, you send a postal survey to 144 people, but only 108 of them respond. What proportion respond? Here, the total number of people is 144 and the number we are interested in, the number who responded, is 108. To find the proportion who responded, we divide 108 by the total 144 as follows:

$$\text{Proportion responding} = \frac{108}{144} = 0.75$$

The percentage responding will be the proportion multiplied by 100:

$$\text{Percentage responding} = \frac{108}{144} \times 100 = 75\%$$

So to calculate a proportion or percentage from some data, the rules are:

$$\text{Proportion} = \frac{\text{number of interest}}{\text{total}}$$
$$\text{Percentage} = \frac{\text{number of interest}}{\text{total}} \times 100$$

Another example would be if you counted that 63 students out of the 94 studying sociology were female. The proportion and percentage of female students are calculated as follows. (Normally you would calculate either the proportion *or* the percentage, not both.)

$$\text{Proportion female} = \frac{63}{94} = 0.6702$$
$$\text{Percentage female} = \frac{63}{94} \times 100 = 67.02\%$$

Watch out, in case you are not given the total. If a friend studying politics told you that there were 82 males and 44 females in his year group, you must first find the total number of students studying politics before calculating the proportion who are male or female:

$$\text{Total students} = 82 + 44 = 126$$

$$\text{Proportion male} = \frac{82}{126} = 0.6508$$

$$\text{Proportion female} = \frac{44}{126} = 0.3492$$

Note that because all the students have been classified as either male or female, the two proportions add up to exactly 1.

Percentage tables

Suppose you have been asked by an environmental charity to find out which methods of transport people use to get to work in two different urban areas. You find some data from the 2011 Census, an extremely useful source of information, and now want to draw a percentage table.

One way of calculating the percentages is shown in Table 2.4. Here the percentages have been calculated separately for each area. Percentages which add up to 100% vertically like this are known as **column percentages**. From this table, the two areas can be easily compared. In Tunbridge Wells, a much higher proportion

Table 2.4 Employees and self-employed residents of Tunbridge Wells and Southampton using different methods of transport to work, 2011 Census (column %)

	Town/city of residence	
Method of transport	Tunbridge Wells	Southampton
Rail	14.14	2.77
Bus, minibus or coach	2.25	9.05
Underground, metro, light rail or tram	0.20	0.10
Pedal cycle	1.12	4.54
Taxi	0.27	0.42
Motorcycle, scooter or moped	0.60	1.04
Driving a car or van	49.47	51.85
Passenger in a car or van	3.82	6.54
Work mainly at or from home	14.35	7.25
On foot	13.39	15.78
Other method of travel to work	0.39	0.64
Total %	100.00	100.00 (99.98)
(Number)	(57,234)	(112,608)

Source: Office for National Statistics, 2013a

of workers travel by train than in Southampton. This is probably because Tunbridge Wells is closer to London and many people commute to the City from there. On the other hand, greater proportions of workers in Southampton travel by bus and pedal cycle than in Tunbridge Wells.

Alternatively the percentages could be calculated as in Table 2.5. These are **row percentages,** which add up to 100% horizontally. Table 2.5 tells us, for example, that of workers travelling on foot, 30.13% live in Tunbridge Wells and 69.87% live in Southampton. This does not really help us to compare transport use in the two areas. The percentages are actually rather easy to misinterpret because the total number of people from Southampton in the table is much greater than the total number from Tunbridge Wells. Most of the row percentages are higher for Southampton purely for this reason, irrespective of transport use.

Table 2.5 Employees and self-employed residents using different forms of transport to work in Tunbridge Wells and Southampton in the 2011 Census (row %)

	Town/city of residence		Total % (number)
Method of transport	Tunbridge Wells	Southampton	
Rail	72.12	27.82	100.00 (11,209)
Bus, minibus or coach	11.20	88.80	100.00 (11,487)
Underground, metro, light rail, tram	49.79	50.21	100.00 (235)
Pedal cycle	11.13	88.87	100.00 (5,758)
Taxi	24.64	75.36	100.00 (629)
Motorcycle, scooter or moped	22.73	77.27	100.00 (1,518)
Driving a car or van	32.66	67.34	100.00 (86,703)
Passenger in a car or van	22.88	77.12	100.00 (9,544)
Work mainly at or from home	50.15	49.85	100.00 (16,380)
On foot	30.13	69.87	100.00 (25,437)
Other method of travel to work	23.46	76.54	100.00 (942)

Source: Office for National Statistics, 2013a

The moral of the story is always to think about which type of percentages you need when constructing a table. Will row or column percentages make the most sense? Which will be more useful for answering the research question? This principle also applies to tables of counts rather than percentages: always add up the totals in the direction which makes most sense. Once you have decided, always make it clear which way the percentages add up by including the 100% totals.

Always ensure that somebody reading the table could find out the original numbers if they wanted to. Do this either by putting the actual numbers in brackets (as in Table 2.2) so they can simply be read from the table, or by including row and/or column totals into the table (as in Tables 2.4 and 2.5) so that somebody with a calculator could work out the actual number in each category. For example, from Table 2.4, 14.14% of Tunbridge Well's workers use rail transport, and 14.14% of 57,234 is 8,093 people. If you do not do this, somebody reading the table will not know if the percentages are out of 1,000 people or out of 10 people and this could put them off using the findings! Remember, the larger the sample size the more likely it is to be representative.

Table 2.6 by Harker (2012), using data from the Department of Education, is another example of a percentage table with column percentages. This enables placements for looked-after children in local authority care in 2007 to be compared with the figures from 2011.

Table 2.6   Children looked after (including adoption and care leavers) at 31 March by types of placement, England, 2007 and 2011

	Percentages[2]	
	2007	2011
Foster placements	70[2]	74
Placed for adoption	5	4
Placement with parents	9	6
Other[1]	3	4
Secure units, children's homes, hostels	11	9
Other residential settings	1	2
Residential schools	2	1
Total %	100	100
(Number)	(60,000)	(64,000)

[1]'Other': placement in the community including living independently and residential employment.
[2]Figures are rounded to the nearest percentage.

Source: Harker, 2012

The information is presented in such a way that it is easy to identify that there are many similarities in terms of the types of placements for looked-after children in 2007 and 2011. Foster placements remained the most common form of placement and expanded slightly over the four-year period, while there was a slight reduction in the percentage of placements with parents. It would be interesting to see to what extent such changes may be a result of a response to high-profile cases such as the death of Baby Peter, which occurred after the 2007 survey results.

Reporting data

Although percentages are easy to work out, they need to be used with caution, as we saw with the alcohol consumption example. For instance, suppose we undertook a survey of 100 students in Sheffield, asking them whether they would describe themselves as religious or not, and 41 out of the 100 stated that they would describe themselves as religious. If these findings were reported in the following fashion it would be problematic:

- 41% of University of Sheffield students described themselves as religious.

How meaningful the claim of 41% actually is depends largely on the size of our sample and our sampling method. There are approximately 27,000 students at the University of Sheffield. If we interviewed 100 students our sample is hardly likely to be representative of the University of Sheffield student population or the student population as a whole. As a result, authoritatively claiming that 41% of University of Sheffield students described themselves as religious would be a little dubious as our data are very limited and unlikely to be representative. This is without exploring how the sample was selected and what we mean by 'religious'. However, if we summarised the results in the following format, we'd be on much safer ground and would not be making claims about our sample (100 University of Sheffield students) which sounded like they were based on the whole University of Sheffield student population (approximately 27,000 students).

- 41% of University of Sheffield students in a sample of 100 students described themselves as religious.

We will discuss issues with reporting data at various stages throughout the book, but it is important to recognise at this early stage that the way data are presented has considerable implications for how they are subsequently reported. So you need to be careful about how you present statistics and also how you discuss them.

Summary

This chapter has introduced you to data and levels of measurement. Without this understanding it is impossible to produce tables effectively. It has shown how there is more to creating an effective table than most people think. This chapter provides clear guidance on the process of producing a meaningful table. This includes the use of headings, the body of the table and notes. In addition, it has covered the construction and use of percentages in tables as well as (briefly) looking at how to report information from tables. For instance, using examples such as the modes of transport in Tunbridge Wells and Southampton, it clearly showed that the way in which data are presented and reported can affect how they are interpreted.

PRACTICE QUESTIONS

2.1 Table 2.7 shows the marital status composition of the British population in 2010 by sex. Construct an appropriate table to see whether there are any differences in marital status between males and females. What does the table show?

(Continued)

(Continued)

Table 2.7 Legal marital status of males and females (all ages) in England and Wales, in thousands, 2010

	Sex	
Marital status	Male	Female
Single[1]	14,119.6	12,561.8
Married[2]	10,504.6	10,589.1
Widowed[3]	720.3	2,419.2
Divorced[4]	1,884.1	2,441.9

[1]People who have never been legally married
[2]People who are currently legally married (including those who are separated)
[3]People who are legally married until the death of their partner, and have subsequently neither remarried nor divorced
[4]People who are legally married but have been legally divorced, or had their marriage annulled, and have not since remarried

Source: Office for National Statistics, 2011

2.2 Table 2.8 by Worcester and Herve from Ipsos MORI (2010) shows voting behaviour by newspaper readership in the 2010 general election.

(a) Construct an appropriate percentage table using these data. Calculate the percentages to the nearest whole number only.

(b) Answer the following questions to check your understanding:

(i) What percentage of the *Sun* readers voted Labour?

(ii) What percentage of the *Guardian* readers voted Conservative?

(iii) What patterns in voting are apparent from the table?

Table 2.8 Voting behaviour by newspaper readership in the 2010 general election

Newspaper read	Party voted for				
	Conservative	Labour	Lib. Dem.	Other party	Total
Mirror	85	315	91	43	534
Daily Telegraph	356	36	96	20	508
Guardian	48	244	196	42	530
Sun	425	276	178	109	988

Source: Worcester and Herve, 2010

THREE
GRAPHS AND CHARTS

Introduction

Producing graphs enables you to learn a lot about a dataset at a glance. Arguably, a graph or chart should be easier for the average person to understand than a table: after all, everybody prefers to look at a pretty picture rather than wade through a table of numbers. The only drawback is that you cannot always display as much data in a graph or chart as in a table, so you need to think very carefully about the audience you are writing for when deciding which format to use. The correct type of graph to use depends on the type of data you have and the question you are addressing. Do not use a particular type of graph just because it looks pretty! For instance, if you have categorical variables then bar charts or pie charts should be used, but if you are working with continuous data then a histogram is the most appropriate form of presentation. It is also worth considering the use of time series data and time-to-an-event data. As with tables, there are a number of presentational guidelines worth sticking to in order to present the data meaningfully. This chapter will introduce you to the process of constructing graphs and charts for different types of variables and how to present them effectively. By the end of the chapter you should be able to:

- Identify when it is appropriate to use different kinds of graphs and charts, including bar charts, line graphs, pie charts and histograms
- Present data in the form of a stem and leaf plot
- Construct ways of presenting time series data and time-to-an-event data
- Understand good practice in the development of different forms of graphs and charts

Graphs for discrete data

You should remember from the previous chapter that discrete variables are those based on categories. The data in Table 3.1 are discrete because they are classified by categories such as 'reducing fraud' and 'making sure payments are fast and accurate', which do not take numeric values. 'Priorities for improving the benefits system' is a nominal variable where there is no inherent order to the attributes.

Table 3.1   Priorities for improving the benefits system, 2010

Priorities for improving the benefits system	First highest priority	First or second highest priority
	%	%
Targeting benefits only at those who really need them	33	53
Rewarding those who work or look for work	18	37
Making sure those who are entitled to money claim it	14	26
Reducing fraud	13	32
Making sure those who save are not penalised	12	26
Providing benefits for those who cannot work	6	14
Making sure payments are fast and accurate	3	7
Number	3,297	3,297

Source: Park et al., 2012

Table 3.1 uses data from a large-scale survey called the British Social Attitudes Survey, which spans a multitude of areas warranting public consideration such as health, crime, marriage, education and benefits. One of the questions it asked was around people's priorities for improving the benefits system in 2010. It provided a number of options in the form of discrete categories. This is a type of nominal variable.

Table 3.2 uses data from an online poll completed by YouGov of 2070 UK adults for the Alzheimer's Society in December 2011 (fieldwork was undertaken between 23 and 29 December 2011) where individuals were asked about their own quality of life and their perceptions of how well people are able to live with dementia. It asks 'how prepared the general public feels UK society is for dealing with people with dementia, breast cancer and diabetes'. This is an ordinal variable given that the level of preparedness is on a sliding scale, but it is non-specific. There is a lot of room for interpretation – the categories are not neat and well bounded. What is 'very prepared' for one person may be 'fairly unprepared' for another. In this sense, the distances between the points on the scale are not equal or well defined.

Table 3.2   How prepared the general public feels UK society is for dealing with people with dementia, breast cancer and diabetes, 2012

	Illness		
	Dementia	Breast cancer	Diabetes
	%	%	%
Very prepared	2	10	12
Fairly prepared	15	51	46
Neither prepared nor unprepared	15	15	14
Fairly unprepared	34	11	15
Very unprepared	26	5	5
Don't know	8	8	8

Source: Alzheimer's Society, 2012

While it is possible to interpret the figures in Tables 3.1 and 3.2 from the tables directly, it is often easier to compare trends by producing graphs or charts. There are a number of ways in which this can be done.

Bar charts

The most straightforward type of graph to produce is a **bar chart**. While the exact frequencies or percentages can be difficult to determine from bar charts, they can be used to summarise visually the distribution of nominal or ordinal data. Remember that we said that nominal data have no mid-points between the categories? You are either in one or the other – never between. So the categories are mutually exclusive and non-continuous. You fit into one category or another. Similarly, remember that we highlighted that the scaling of ordinal data is often difficult to determine and as a result ordinal data should not be treated as continuous? Look at the gaps between the bars – these gaps indicate that the data we are working with are not continuous. This is why we use bar charts to describe nominal and ordinal data.

It's also worth remembering that when you are using bar charts, too many categories make the graph very difficult to read – we wouldn't want many more categories than the ones we currently have in the following examples.

Figure 3.1 shows the British Social Attitudes data from Table 3.1 presented as a bar chart. The chart shows clearly that, of the seven possible priorities for benefit reform, targeting benefits only at those who really need them is the most common highest priority. It is easier to see this at a glance from the graph than it was from the original table, so the graph has achieved its aim.

Similarly, Figure 3.2 presents the information about how prepared the general public feels UK society is for dealing with people with dementia, an ordinal variable, on a bar chart. Again it is easy to see that there are considerable percentages of people who feel the UK is fairly or very unprepared for dealing with people with dementia.

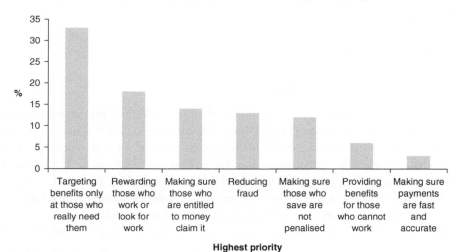

Figure 3.1   A bar chart showing the percentage of people's highest priorities for improving the benefits system, 2010

Source: Park et al., 2012

Note that the bars in a bar chart are always rectangular and are all the same width. The height of each bar is proportional to the number or percentage in each category.

Also notice the ordering of the bars. There is no hard and fast rule here, but a common practice is to order them from shortest to tallest or tallest to shortest, as in Figure 3.1. Sometimes you may want to order the bars like this, but leave a category such as 'none' or 'other' at the end. In some cases, the categories themselves may have a natural order: for example, social class or educational qualifications should be arranged in a logical order from highest to lowest or vice versa, irrespective of the height of the bars. This has been done in Figure 3.2, where the bars are ordered from 'very prepared' to 'very unprepared', with the 'don't know' category at the end.

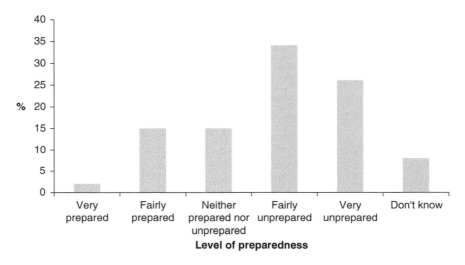

Figure 3.2   A bar chart showing how prepared the general public feels UK society is for dealing with people with dementia, 2012

Source: Alzheimer's Society, 2012

Multiple bar charts

The bar chart showing the percentages provided for the highest priorities for improving the benefits system in the British Social Attitudes Survey 2010 is fine, but we may also be interested in showing both 'highest priorities' and 'first or second priorities' as we had done in the initial frequency table. This can be done on a multiple bar chart, as in Figure 3.3, where the bars for 'highest priority' and 'first or second priority' stand side by side. A key (legend) is important now, to identify which bar represents the 'highest priority' and which represents 'first or second priority'.

The multiple bar chart makes it easy to see that both 'targeting benefits only at those who really need them' and 'rewarding those who work or look for work' are not only the most represented highest priorities identified but also the most common first and second priorities. Figures such as these are often used to justify support for policy changes.

This figure only has two sets of bars and is fairly clear. However, you should try to avoid putting more than three or four sets of bars on a multiple bar chart, as the chart becomes cluttered and difficult to interpret.

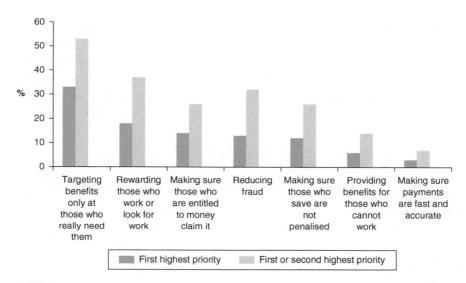

Figure 3.3 A bar chart showing the percentage of people's first highest priorities and first or second highest priorities for improving the benefits system, 2010

Source: Park *et al.*, 2012

Stacked bar charts

Bar charts and multiple bar charts are suitable for working with categories which make up part of a whole (i.e. add up to 100%) or where all the categories are separate and have no numerical relationship to each other as they compare the size of each category with the size of other categories. However, stacked bars and pie charts can only be used for data where the categories make up part of a whole. They make it easier to see what proportion one category is of the whole.

To draw a stacked bar chart by hand, the **cumulative percentages** must be calculated. The following example shows the calculation of cumulative percentages for 'how prepared the general public feels UK society is for dealing with people with dementia'. All it involves is some simple addition, starting with the first percentage and then adding each subsequent percentage to the total. You should end up with 100% as the final cumulative percentage; if not, something has gone wrong!

The data are then ready to be presented on a chart as in Figure 3.4. The categories should be ordered appropriately as before. Either the labels can be included in a key (legend) as in Figure 3.4, or each section can be labelled at the side of one of the bars, providing it is clear for both bars which part is which. Always use the same colour or shading for the same category in different bars.

Having the three columns side by side makes it easy to identify any differences. In Figure 3.4, you can see that the Alzheimer's Society's (2012) findings about 'how prepared the general public feels UK society is for dealing with people with dementia, cancer or diabetes' indicate considerable differences. The levels of preparedness for 'dementia' are substantially lower than for 'breast cancer' and 'diabetes', with 61% stating that UK society is 'very prepared' or 'fairly prepared' for dealing with people with 'breast cancer' and 58% for 'diabetes', compared to only 17% for 'dementia'. This could be because younger people are not very aware of it as it is mainly experienced at older ages. The last sentence is purely speculation made after observing the data; take care to distinguish between factual observations about a graph and your own suggestions for any patterns or trends that you can see.

	%	Calculation	Cumulative %
Very prepared	2	2	2
Fairly prepared	15	2 + 15	17
Neither prepared nor unprepared	15	17 + 15	32
Fairly unprepared	34	32 + 34	66
Very unprepared	26	66 + 26	92
Don't know	8	92 + 8	100

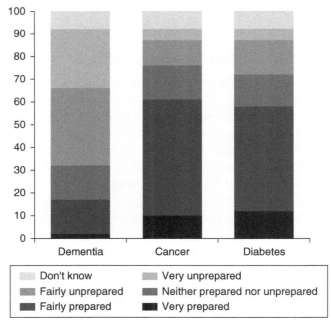

Figure 3.4   **A stacked bar chart showing how prepared the general public feels UK society is for dealing with people with dementia, breast cancer and diabetes, 2012**

Source: Alzheimer's Society, 2012

Pie charts

Pie charts must only be used when the categories of data make up part of a whole, otherwise the chart will be meaningless. The data must be in the form of either percentages or proportions, so if they are not, the first step is to convert them into one of those forms.

A pie chart presents the categories of data as parts of a circle or 'slices of a pie'. A circle contains 360 degrees (360°), so to convert a proportion to degrees, simply multiply it by 360. To convert a percentage to degrees, first divide by 100 to convert it to a proportion and then multiply by 360. The following tabulation shows how the percentages for 'how prepared the general public feel UK society is for dealing with people with dementia' are converted into degrees. If the calculation has been done correctly the slices should add up to 360°, give or take a degree for rounding. The figures have only been calculated to the nearest degree because it is difficult to measure much more accurately than this with a protractor if you are drawing the chart by hand.

	%	Proportion (% ÷ 100)	Degrees (proportion × 360)
Very prepared	2	0.02	7
Fairly prepared	15	0.15	54
Neither prepared nor unprepared	15	0.15	54
Fairly unprepared	34	0.34	122
Very unprepared	26	0.26	94
Don't know	8	0.08	29
Total			360

Then the pie chart can be drawn as in Figure 3.5. It is usual to start measuring your angles from a vertical line at the top of the circle, but some people start from a horizontal line on the right.

Include a key to show what the different colours/shadings in each 'slice' represent. If you are drawing several pie charts on the same subject (such as expenditure in several different countries), always put the categories in the same order on each pie chart: this avoids confusion. Sometimes the circles may be drawn in different sizes in proportion to the size of the total, for example, the total number of people in the sample.

While pie charts are excellent at helping an audience gain a sense for the distribution quickly, they do have some limitations. Indeed, deciphering wedges in a pie chart can be more difficult than comparing the heights of bars. Angles are typically harder to compare than lengths, so pie charts are not always that great for comparing different quantities – especially where they are similar. This is why it is important to provide the numbers too.

Like the stacked bar chart in Figure 3.4, the pie charts in Figure 3.5 clearly show that there are considerably lower levels of preparedness for dementia than breast cancer and diabetes in the UK. Whether you use stacked bars or pie charts comes down to personal preference, but try to use the type which shows the data in question more clearly.

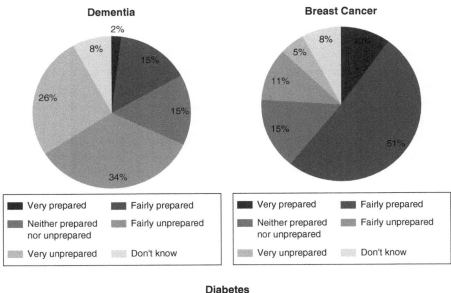

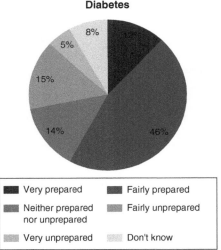

Figure 3.5 Pie charts showing how prepared the general public feels UK society is for dealing with people with dementia, breast cancer and diabetes, 2012

Source: Alzheimer's Society, 2012

Rules for good graphs

Before moving on to look at other types of graph, let's just summarise some of the ground rules for producing an excellent graph of any type. It is also worth noting that many of the characteristics highlighted as being important in tables are also important in graphs, so take a look back at the guidance in the previous chapter on tables too.

- *Always use the correct type of graph for your data.* Ask yourself whether the data are discrete or continuous, or whether you have data showing change over time. Then use an appropriate graph type. Do not choose your favourite type of graph and fit the data into it!
- *Label the graph clearly.* All graphs should have a clear title explaining what or who the data refer to and when and where they refer to. For example, 'Infant mortality rates in Malawi, 1997–2012' is far more informative than 'A graph of IM rates'. Both the *X* (horizontal) and *Y* (vertical) axes of the graph should also have titles, not forgetting the units of measurement.
- *The graph itself should be neat and clear.* Apart from obvious scruffiness, avoid cluttering up a graph with too many bars or lines. Always use the same colours or shadings for the same categories when you have more than one pie chart or stacked bar. Also limit the number of different shadings used in a single graphic. If different shadings are utilised to distinguish bars of a graph, choose shadings that are distinct. Include a key (legend) where appropriate. The *Publication Manual* of the APA (2010) states the need to consider the *weight* (i.e. size, density) of each part of a graph in relation to every other part, making the most important parts the most prominent. For example, curves on line graphs should be bolder than axis labels, which should be bolder than the axes.
- *Include the source.* You will normally know where the data came from (perhaps a book or a survey). Put a note below the graph to this effect; for example 'Source: British Social Attitudes Survey, 2013'. Or include the source in the title.

Using a computer

Once you have learnt the basic steps, it is relatively easy to produce professional looking graphs on a spreadsheet or statistical package.

Do not be deceived! *It is also very easy to produce a completely meaningless graph on a computer.* Take as much care with a computer graph as you would with a hand-drawn one. Just because a graph looks neat and impressive in print does not necessarily mean that you have chosen the correct type of graph for the data or labelled the graph properly.

Graphs for continuous data

95

On seeing this number floating aimlessly in mid-air, you would naturally wonder what it is. Out of context, the number 95 could mean any number of things.

In fact, 95 is the percentage of births attended by trained health personnel (such as midwives or doctors) in Tunisia between 2005 and 2011 according to the World Health Organization (2012).

This is interesting, but it simply leads to more questions, such as:

- How many African countries have a higher percentage of births attended by trained health personnel than Tunisia? How many countries have a lower percentage?
- Is there a big variation between countries or do they all have a similar percentage of births attended by trained health personnel?
- What percentages of African countries have 50% or less of their births attended by trained health personnel?

These are questions which ask about the **distribution** of the percentage of births attended by trained health personnel in African countries.

Table 3.3 gives some raw data for some African countries. It is not very clear: it's just a jumble of numbers! With more data it would get even murkier! What can we do about it? It's worth noting that the percentage of births attended by trained health personnel in African countries represents a ratio variable as it has an absolute zero point and there is an equal distance between each percentage point, for example 18% and 19% or 35% and 36%.

Table 3.3 Percentage of births attended by trained health personnel in 43 African countries for which figures are available, 2005–11

95	74	62	55	82	49	78	46	81
95	95	79	100	18	42	67	10	44
34	60	71	41	83	57	80	52	55
46	44	44	71	49	57	100	74	69
65	31	9	23	49	47	66		

Source: World Health Organization, 2012

One way of making the data easier to look at is to use an **array**. This just means putting the data in numerical order, as in Table 3.4. Having ordered the data, it is possible to see that Tunisia actually has one of the highest percentages of births attended by trained personnel among the 43 African countries included. So Tunisia is doing relatively well with respect to maternal care at childbirth.

Table 3.4 An array of the percentage of births attended by trained health personnel in 43 African countries for which figures are available, 2005–11

9	10	18	23	31	34	41	42	44
44	44	46	46	47	49	49	49	52
55	55	57	57	60	62	65	66	67
69	71	71	74	74	78	79	80	81
82	83	95	95	95	100	100		

Source: World Health Organization, 2012

Alternatively, you could use a **complete frequency distribution**. As Table 3.5 shows, this involves listing every value and the 'frequency' or number of times it occurs. This helps us to see which values occur more than once, but otherwise is not much more useful than an array.

If there were more data in the table it would become too large and unmanageable. The solution is to **group** the data. The table then becomes an **abridged frequency table**, as in Table 3.6. This table takes up much less space and tells us much more about the distribution of the data. For example, we now know

Table 3.5 A complete frequency distribution of the percentage of births attended by trained health personnel in 43 African countries for which figures are available, 2005–10

% births attended by trained health personnel	Frequency (number of countries)	% births attended by trained health personnel	Frequency (number of countries)	% births attended by trained health personnel	Frequency (number of countries)
9	1	49	3	78	1
10	1	55	2	79	1
18	1	57	2	80	1
23	1	60	1	81	1
31	1	62	1	82	1
34	2	65	1	83	1
41	1	66	1	95	3
42	1	67	1	100	2
44	3	69	1		
46	2	71	2		
47	1	74	2		
Total					43

Source: World Health Organization, 2012

Table 3.6 An abridged frequency table of the percentage of births attended by trained health personnel in 43 African countries for which figures are available, 2005–10

% births attended by trained health personnel	Frequency (number of countries)
1–10	2
11–20	1
21–30	1
31–40	2
41–50	11
51–60	6
61–70	5
71–80	7
81–90	3
91–100	5
Total	43

Source: World Health Organization, 2012

that in two countries, only between 1% and 10% of births are attended by trained health personnel. The interval with the most countries in it is the 41–50% group, with 11 countries. There are five countries with between 91% and 100% of births attended by trained health personnel, and Tunisia is one of these.

Steps for creating an abridged frequency table

Step 1: find the range of the distribution and determine the width of each class interval

The percentages in the example range from 9 to 100. Therefore it might be appropriate to start at 1 and go up to 100.

The width of each class interval should be chosen bearing in mind the range and the size of the dataset. You do not want too many or too few observations in a class. Using class intervals of 1–49 and 50–100 will only give you two classes and there will be so many observations in each class that the distribution will be hidden. On the other hand, using class intervals of 7–9, 10–12, 13–15, 16–18 and so on will give you too many classes, which is equally unhelpful.

There is no right or wrong width as such, but all classes must be the same width and the width chosen should be *sensible*! Ideally you might want five to ten classes if you have a reasonable number of observations. Table 3.6 has ten class intervals of width 10.

Step 2: list the class intervals, starting at the bottom

Make sure that the numbers do not overlap, in other words use 1–10, 11–20, ..., not 0–10, 10–20, ... Do not use obscure numbers for class boundaries. The classes 0–15.6, 15.7–31.2, ... will only confuse people!

Step 3: find the frequency in each class

You might want to go through the data, crossing off each observation and putting a tally mark in the appropriate class. When all the observations have been crossed off, add up the tally marks to find the total number or frequency in each class. Always add up the frequencies for each class to check that they add up correctly to the total number of observations.

Now we have completed a frequency distribution, but it still doesn't show us the *percentage* of countries in each category. For example, we might want to know what percentage of countries have fewer than 50% of births attended by a trained person. In order to work this out we must work out the percentages and cumulative percentages. The calculations are fairly straightforward. The percentages are worked out in the usual way. For example, the percentage of countries with between 1% and 10% of births attended is calculated as follows (to two decimal places):

$$\text{Percentage} = \frac{2}{43} \times 100 = 4.65\%$$

Once the percentage in each group has been calculated, always add up the percentages to make sure that they add up to 100% in total, as in Table 3.7. (Answers of 100.01 or 99.99 are perfectly acceptable due to rounding!)

To calculate the cumulative percentages, simply add up the percentages as you go down the table. The top cumulative percentage (for the 1–10 group) will be the same as the first percentage, 4.65%. The cumulative percentage for the 11–20 group is 4.65 + 2.33 = 6.98. For the 21–30 group the cumulative percentage will be 6.98 + 2.33 = 9.31. So, each time, the percentage of the group is added to the previous cumulative percentage. The final cumulative percentage should equal approximately 100%.

Table 3.7 Percentages and cumulative percentages for the percentage of births attended by trained health personnel in 43 African countries, 2005–10

% of births attended by trained health personnel	Frequency (number of countries)	% of countries	Cumulative %
1–10	2	4.65	4.65
11–20	1	2.33	6.98
21–30	1	2.33	9.31
31–40	2	4.65	13.96
41–50	11	25.58	39.54
51–60	6	13.95	53.49
61–70	5	11.63	65.12
71–80	7	16.28	81.40
81–90	3	6.98	88.38
91–100	5	11.63	100.01[1]
Total	43	100.01	100.01

[1]Percentages do not add up to exactly 100% due to rounding.

Source: World Health Organization, 2012

The cumulative percentage is the percentage of scores which lie below the upper limit of a particular interval. So, for example, 9.31% of the countries have 30% or fewer births attended by a health professional (because 30 is the top of this group).

Going back to our original question, what percentage of the countries have 50% or fewer births attended by trained health personnel? The answer is 39.54% of the countries (39.54% have 50% or fewer births attended).

Note that it is also possible to work out the **cumulative frequency** for each group by adding up the frequencies going down the table instead of the percentages. You could also work out the proportion of countries in each group; this is known as the **relative frequency**. Which of these methods you use depends on the questions you want to answer. If you had wanted to know the *number* of countries with 50% or fewer births attended by trained health personnel, the cumulative frequency would be more useful than the cumulative percentage.

Stem and leaf plots

A frequency distribution can be made more visually appealing by turning it into a **stem and leaf plot**. Table 3.8 shows the percentage of males and females who were literate in 44 African countries in 2005–10.

What can we discover about female literacy in Africa from these data? In their current state, not a lot! A jumble of numbers is not very useful, so to make them easier to interpret, an array or frequency table could be drawn up. Alternatively, the data could be presented as a stem and leaf plot, as in Figure 3.6.

The column to the left of the line is known as the **stem**, while the other numbers to the right of the line are the **leaves**. The stem represents the 'tens' and the leaves the 'digits'. For example, the row beginning with '5' has the digits '0', '1', '7', '7', '8' and '8'. Therefore we know there are observations of 50, 51, 57, 57, 58 and 58 among the data.

Table 3.8 Literacy rate among males and females aged 15 and over in 44 African countries, 2005–10 (%)

Males

81	83	55	84	37	73	79	69	49	65
77	80	79	49	92	60	73	52	61	91
83	65	96	67	81	43	65	91	69	71
89	43	72	75	62	54	91	80	88	86
83	79	81	95						

Females

64	58	30	85	22	62	63	43	24	47
57	64	58	29	85	40	61	30	41	84
96	57	83	62	68	20	51	86	44	43
88	15	50	68	39	31	87	62	87	71
65	67	62	90						

Source: UNESCO Institute for Statistics, 2013

```
1 | 5
2 | 0 2 4 9
3 | 0 0 1 9
4 | 0 1 3 3 4 7
5 | 0 1 7 7 8 8
6 | 1 2 2 2 3 4 4 5 6 7 8 8
7 | 1
8 | 3 4 5 5 6 7 7 8
9 | 0 6
```

Figure 3.6 A stem and leaf plot showing literacy rate among females aged 15 and over in 44 African countries for which figures are available, 2005–2010 (%)

Source: UNESCO Institute for Statistics, 2013

Steps for drawing a stem and leaf plot

Mentally separate each digit into a stem and a leaf, for example 22 = stem 2, leaf 2. Note that stems can have as many digits as you like, but leaves can only have one digit. For larger numbers, such as 176, the stem would be 17 and the leaf 6.

List the stems, increasing as you move downwards. If you feel that you want intervals narrower than tens, another possibility is to divide the stems into two, where for example the stem '0' is for the leaves 0 to 4 and '0*' is for the leaves 5 to 9. Figure 3.7 shows part of the data in the example presented in this way.

Add the leaves to the diagram, in numerical order, spacing the numbers evenly. Figure 3.8 shows how *not* to draw a stem and leaf plot. It is important to space the numbers evenly so that the number of observations in each row can be easily compared.

```
4  | 0 1 3 3 4
4* | 7
5  | 0 1
5* | 7 7 8 8
```

Figure 3.7 Part of the stem and leaf plot in Figure 3.6 with stems divided into two

```
1 | 5
2 | 0 9 2 4
3 | 0 1 9 0
4 | 7  0 14 3 3
5 | 078  7 1
```

Figure 3.8 How *not* to draw a stem and leaf plot

Interpreting a stem and leaf plot

Look first at the shape of a stem and leaf plot. Are the data concentrated around the middle of the distribution with symmetric tails (a **normal distribution**), or is the distribution **skewed** with most of the data concentrated at one end of the plot? In Figure 3.6, most of the data lie just above the middle of the plot (12 countries with 60–69% literacy), with fewer countries having very high or very low values.

It is worth thinking about where the 'average' value might lie: where is the middle of the distribution? In Figure 3.6 we might guess that the average value is in the 60s. Finally, check whether there are any **outliers** (points all by themselves at one end of the range). If there are any, it might be a good idea to check that the values have been inputted correctly.

What does Figure 3.6 tell you about female literacy in Africa in 2005–10?

Back-to-back stem and leaf plots

So far we have only examined the data for females. It would be interesting to compare female literacy with male literacy in the 44 countries to see whether there are any gender differences. In order to compare two different distributions, two stem and leaf plots can be drawn next to each other, sharing the same stem. This is known as a **back-to-back stem and leaf plot**. Figure 3.9 presents the data for both males and females. It is now very easy to compare the shapes of the two different distributions.

What does the plot tell us about gender differences in literacy in Africa in 2005–10? Straight away we can see that male literacy tends to be higher than female literacy because there are many more countries with literacy above 70%

Males			Females
	1	5	
	2	0 2 4 9	
7	3	0 0 1 9	
9 9 3 3	4	0 1 3 3 4 7	
5 4 2	5	0 1 7 7 8 8	
9 9 7 5 5 5 2 1 0	6	1 2 2 2 3 4 4 5 6 7 8 8	
9 9 9 7 5 3 3 2 1	7	1	
9 8 6 4 3 3 3 1 1 1 0 0	8	3 4 5 5 6 7 7 8	
6 5 2 1 1 1	9	0 6	

Figure 3.9 A back-to-back stem and leaf plot showing literacy rate among males and females aged 15 and over in 44 African countries for which figures are available, 2005–10 (%)

Source: UNESCO Institute for Statistics, 2013

for males than there are for females. Correspondingly there are no countries with male literacy below 30%, while there are five countries with female literacy below 30%. The average male literacy rate is likely to be higher than the average female literacy rate.

Advantages and disadvantages of stem and leaf plots

The unique advantage of a stem and leaf plot is that it displays all the data. By looking at the plot you know the exact values of all the observations in the dataset, so no accuracy is lost. Back-to-back stem and leaf plots are also particularly good for comparing two distributions side by side.

However, with a very large sample, producing a stem and leaf plot would be very time-consuming and there would simply be too many data to look at all at once. With a large sample, a histogram is a better choice for data presentation.

Histograms

A **histogram** is similar to a bar chart, but is used for interval and ratio level variables. With a histogram, the width of the bar is important, since it is the total area under the bar that represents the proportion of the phenomenon accounted for by each category, not just the height. A set of continuous data is divided up into groups, the frequencies in the groups are found, and a histogram is produced by drawing vertical bars, without gaps between them, whose areas are proportional to the frequencies in the groups. It shows you how often different values occur, how much spread or variability there is among the values and which values are most typical for the data. So in this case, you can display the data in Table 3.8 for the ratio variable 'the percentage of males literate in 44 African countries' as a histogram.

A histogram is a good method of representation if you want to illustrate a set of data in a simple easy-to-read manner. It is also good at indicating outliers – extreme values in a distribution. A histogram should be used for data that are continuous or have been measured on a continuous number scale. For example, you could use a histogram to show the number of live births by age of the mother or the number of years in a job.

Histograms display only the number (or percentage) of observations that fall into each interval, not the actual data in the way that a stem and leaf plot does. In order to plot the data for male literacy on to a histogram, it is useful first to construct an abridged frequency table. Bear in mind that the number of intervals in the table will correspond to the number of bars on the graph. Table 3.9 shows an abridged frequency table for male literacy with intervals of width 10. The histogram can then be constructed as in Figure 3.10. It will be exactly the same shape as the male distribution in the stem and leaf plot in Figure 3.9.

Table 3.9 An abridged frequency table showing the percentage of males literate in 44 African countries in 2005–10

% of males literate	Frequency (number of countries)
0–9	0
10–19	0
20–29	0
30–39	1
40–49	4
50–59	3
60–69	9
70–79	9
80–89	12
90–99	6
Total	44

Source: UNESCO Institute for Statistics, 2013

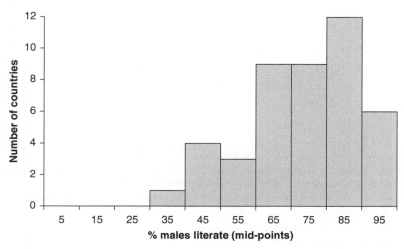

Figure 3.10 Histogram showing literacy rate among males aged over 15 years in 44 African countries for which figures are available, 2005–10

Source: UNESCO Institute for Statistics, 2013

Note that because the data are continuous rather than in discrete categories, the bars of a histogram must touch each other. The height of each bar represents the frequency (number of countries) in that interval, and the width of each bar must be proportional to the width of the interval on the continuous scale. In this case, all the intervals are the same width, so the bars are also the same width.

The way in which the horizontal axis is labelled can vary. The mid-point of the interval can be used in the middle of the bar, as in Figure 3.10, or the bottom of the interval (0, 10, 20, …) can be used at the bottom left of each bar. Alternatively, you can label the bars '0–9', '10–19', and so on.

This example uses the *number* of countries on the *Y* axis, but histograms can also use percentages. This is particularly useful if you are comparing two or more distributions with different numbers of observations because the graphs can still be drawn on the same scale.

The data in Table 3.10 have been converted from frequencies into percentages. The figures were obtained from Labour Force Survey data. Note that 'number of children' is one of those awkward variables which is difficult to classify as either continuous or discrete. The scale from 1 to 10 is clearly ordered and meaningful, but on the other hand it is not possible to have 2.4 or 3.1 children, so the scale is not truly continuous. For now, we will treat the variable as continuous. The data can then be plotted on a histogram, as in Figure 3.11.

Table 3.10 Families with dependent children by number of dependent children in the UK in the 2012 Labour Force Survey

Number of dependent children[1]	Number of families (in thousands)[2]	% of families
1	3663	47.33
2	2983	38.54
3	829	10.71
4	199	2.57
5	50	0.64
6	11	0.14
7	3	0.04
8	1	0.01
9	0[3]	0.00
10	0[3]	0.00
Total	7739	99.98[4]

[1]Dependent children are those living with their parent(s) and either (a) aged under 16, or (b) aged 16 to 18 in full-time education, excluding children aged 16 to 18 who have a spouse, partner or child living in the household.

[2]A family is a married, civil partnered or cohabiting couple with or without children, or a lone parent with at least one child.

[3]These represent fewer than 500 families so have been rounded down to zero.

[4]Percentages do not add up to exactly 100% due to rounding.

Source: Office for National Statistics, 2013b

What is the most common family size among these women? The histogram clearly shows that nearly half (47%) of the families with dependent children in the

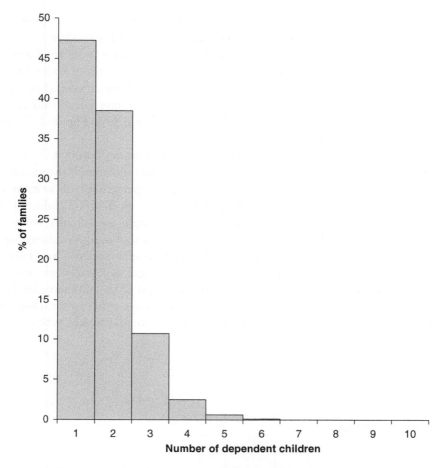

Figure 3.11 Histogram showing families with dependent children by number of dependent children in the UK in the 2012 Labour Force Survey

Source: Office for National Statistics, 2013b

Labour Force Survey, 2012, had only one dependent child living with them. Only a very small percentage of the sample had more than five dependent children in the family. In fact, these percentages are so small that they are extremely difficult to see on the histogram!

In light of this, we might want to combine the very small percentages of families with four or more dependent children into one larger category. If the number of families with 4, 5, 6, 7, 8, 9 and 10 dependent children are added together and converted into a percentage, the result is that 3.4% of families with dependent children have four or more dependent children living with them. The new histogram could then be plotted as in Figure 3.12.

Can you see anything wrong with the histogram in Figure 3.12? Remember that the width of a bar should match the width of the interval. The interval 4–10 has a width of seven children (it includes seven different possible numbers of children), while the other bars only have a width of one child. One solution would be to label the X axis from 0 to 10 as before and then make the bar for the 4–10 group seven bars wide. However, if you do this, you *must* adjust the height of the bar.

The total size of each bar in a histogram is proportionate to its total area, not just its height. Therefore in order to make the 4–10 bar seven times wider, we must also make it seven times shorter! What we are really doing is averaging out the percentage over the seven groups. If 3.4% of the women have 4–10 children, then the height of the bar for this group will be 3.4% ÷ 7 = 0.49%. Figure 3.13 shows the correctly drawn histogram.

Note that this averaging process involves making the assumption that the percentage of women with each number of children is uniformly distributed throughout the large interval of 4–10 children (in other words, the percentage of women with 4 children is the same as the percentage with 5, 6, 7, 8, 9 or 10 children). This assumption may not always be sensible. In this example we actually know how many women have each number of children and know that the percentages are not equal. Figure 3.11 is still the best histogram because no information is lost. However, sometimes the data will be grouped when you obtain them, so you have no choice but to make this assumption and adjust the width and height of the bars accordingly. You should always make it clear when bars are for averaged values, either by including a note to that effect, as in Figure 3.13, or by not dividing the wider bars with vertical lines.

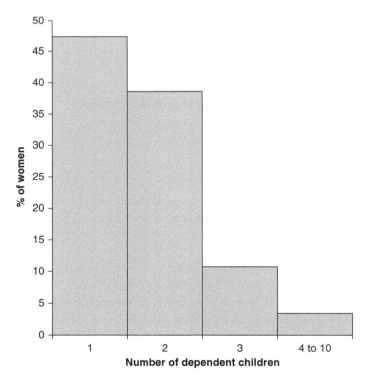

Figure 3.12   **An incorrectly drawn histogram of families with dependent children by number of dependent children in the UK in the 2012 Labour Force Survey**

Source: Office for National Statistics, 2013b

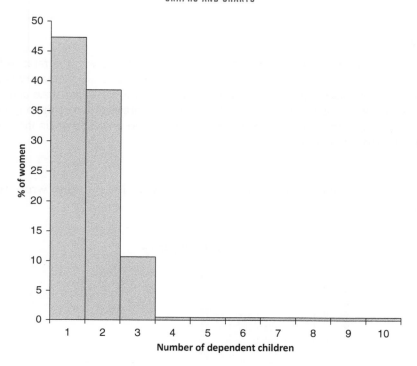

Figure 3.13 **A correctly drawn histogram of families with dependent children by number of dependent children in the UK in the 2012 Labour Force Survey**

¹The figures for 4–10 dependent children have been averaged out over the seven groups
Source: Office National Statistics, 2013b

EXAMPLE: USING HISTOGRAMS TO COMPARE DISTRIBUTIONS

In 1787–8, the famous *Federalist* papers were published in the USA to try to persuade the citizens of New York to ratify the Constitution. Political historians have spent much time trying to determine the authorship of the 85 papers. They agree about the authorship of 73 of the papers, but the remaining 12 papers could have been written by either of two famous political writers, Hamilton and Madison. The political content of the papers does not provide convincing evidence either way as both authors used similar arguments and both changed their political views later in life.

Using histograms, it is possible to throw some light on the question. Different authors tend to have different writing styles and to use non-contextual words such as 'by', 'to' and 'from' to different extents. By examining the frequency of usage of such words in known works by Hamilton and Madison and comparing this to the frequency of usage in the 12 disputed papers, it may be possible to obtain more evidence about the authorship of the disputed papers.

Mosteller and Wallace (1964) used this approach to solve the dispute. They calculated use rates per 1000 words for various words in Hamilton's papers, Madison's papers and the 12 disputed papers. The results for each word were then plotted on

(Continued)

43

(Continued)

histograms and the rates for the disputed papers compared with the rates in the papers of known authorship. Figures 3.14 and 3.15 show their results for the words 'by' and 'to'. Note that the histograms have all been drawn to the same scale and use percentages to make comparisons easier. If you are drawing two or more graphs which you want to compare to each other, always try to present them in the same format and on the same page like this.

The evidence points strongly to one author. Who do you think wrote the 12 disputed papers?

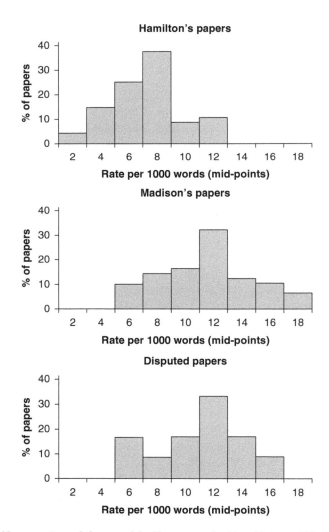

Figure 3.14   Usage rates of the word 'by' in papers by Hamilton and Madison and the 12 disputed *Federalist* papers

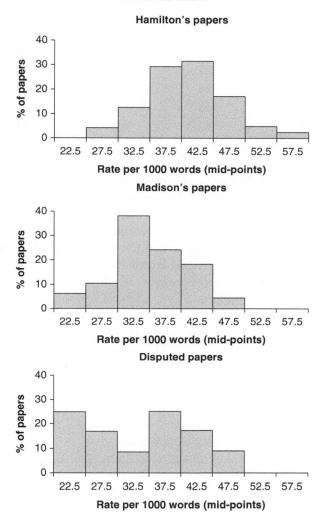

Figure 3.15   Usage rates of the word 'to' in papers by Hamilton and Madison and the 12 disputed *Federalist* papers

Line plots

Line plots are useful for small datasets. They are not generally used for final presentation, but can be drawn up quickly to give you an idea of the distribution of your data.

Suppose a lecturer has taught a group of students on a course. She has just marked all the exam papers and coursework, and wants to quickly get a feel for how the students performed in each.

The ordered results from the exam itself are shown in Table 3.11. Various methods could be used to present these data, including a grouped frequency table, a stem and leaf plot or a histogram. However, the lecturer does not want to spend a lot of time presenting the results, so a line plot would be suitable for simply exploring the data.

Table 3.11   Results for 30 students from an exam (%)

35	40	42	43	44	45
45	46	49	50	50	50
51	52	53	53	54	55
55	58	58	58	58	59
60	60	61	64	64	65

Source: Authors' data

On a line plot, each data value is represented as a blob on a scaled horizontal line. Data of the same value or too close together to be distinguishable are placed in even stacks, as shown in Figure 3.16.

It is easy to see from the line plot how the students performed in the exam. Most of the marks are in the 40s and 50s, with a few marks in the 60s and one mark in the 30s. The students have not done outstandingly well in the exam. Let's see whether they did better in the coursework.

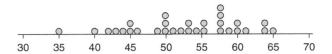

Figure 3.16   A line plot showing the exam results of 30 students

Table 3.12   Results from 30 students' coursework (%)

48	49	50	51	54	55
59	60	60	60	61	61
62	63	64	65	65	66
67	67	67	67	70	70
71	71	74	77	78	80

Source: Authors' data

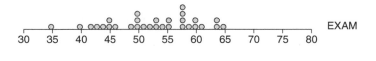

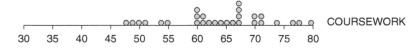

Figure 3.17   A parallel line plot showing the exam and coursework results of 30 students

Table 3.12 shows the students' coursework marks. If the lecturer wants to compare the distribution of exam marks with that of coursework marks, she can draw a **parallel line plot**, as in Figure 3.17. The two scales should be identical on a parallel line plot to enable a straight comparison.

What conclusion might the lecturer draw from Figure 3.17 about the relative performance of her students in the exam and the coursework?

Graphs for time series data

A dataset which shows changes in a variable over time is known as a **time series**. For such data, it is best to use a **line graph**. Numbers, percentages or proportions can all be used on a line graph. On any graph, the horizontal axis is known as the **X axis** and the vertical axis as the **Y axis**. On a line graph, time should always be measured on the X axis.

Table 3.13 shows changes in the percentage of British women aged 16–49 using five different methods of contraception. The data come from Lader (2009), using figures from the Opinions Survey, and we have used information from every two years: 2000/01, 2002/03, 2004/05, 2006/07 and 2008/09.

Table 3.13 Percentage of women in Great Britain aged 16–49 using five methods of contraception, 2000/01–2008/09

	Year				
Method[1]	2000/01	2002/03	2004/05	2006/07	2008/09
Pill	25	25	25	27	25
Intra-uterine device (IUD)	5	5	4	4	6
Male condom	21	20	22	22	25
Other non-surgical	10	10	13	14	15
Male/female sterilisation	22	23	22	20	17

[1] Percentages do not add up to 100% for each year because some women used more than one method and others were not using any method.

Source: Lader, 2009, using figures from the Office for National Statistics Opinions Survey

Figures 3.18 and 3.19 show changes over time in male condom and IUD use. What do the graphs tell you about the trends in the use of the two methods?

Figure 3.18 appears to show a slight increase in the percentage of women using male condoms between 2000/01 and 2008/09, with 25% of women using this form of contraception in 2008/09, a 4 percentage point increase on 2000/01. Incidentally this is nearly double the 13% seen in 1986. We might speculate that male condom use has probably increased due to health promotion campaigns linked to prevention of sexually transmitted diseases. Figure 3.19 suggests that IUD use has increased slightly between 2000/01 and 2008/09 (by 1 percentage point).

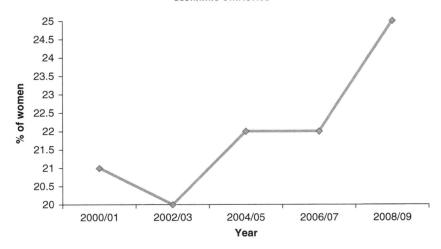

Figure 3.18 Percentage of women in Great Britain aged 16–49 using male condoms, 2000/01–2008/09

Source: Lader, 2009, using figures from the Opinions Survey

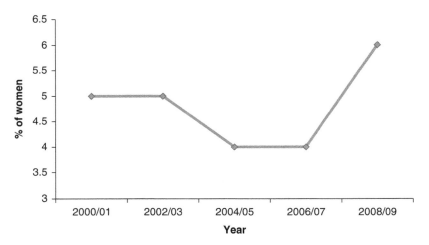

Figure 3.19 Percentage of women in Great Britain aged 16–49 using an IUD, 2000/01–2008/09

Source: Lader, 2009, using figures from the Opinions Survey

It is worth noting in both graphs that the *Y* axis scales do not start at zero. This doesn't make the graphs inaccurate, but it does make them misleading. The vertical scales are deliberately designed so that the increases and decreases appear as large as possible: this clever trick is often used in advertising. For a graph which does not mislead, you should preferably start at zero on the *Y* axis or, if this is not appropriate, at least make it clear to the reader that it does not start at zero by including a break at the bottom of the axis.

A good line graph should show the trend clearly at a glance, without exaggerating or minimising it.

48

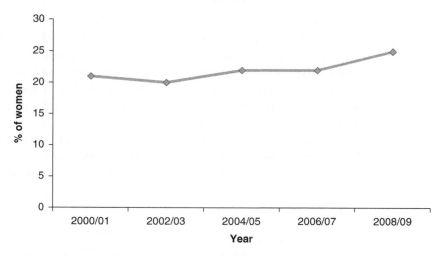

Figure 3.20   Percentage of women in Great Britain aged 16–49 using male condoms, 2000/01–2008/09 (with the Y axis starting at zero)

Source: Lader, 2009, using figures from the Opinions Survey

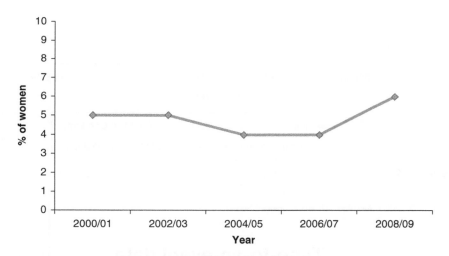

Figure 3.21   Percentage of women in Great Britain aged 16–49 using an IUD, 2000/01–2008/09 (with the Y axis starting at zero)

Source: Lader, 2009, using figures from the Opinions Survey

Figures 3.20 and 3.21 show the trends in the percentage of women using male condoms or an IUD in a less misleading way. The graphs still show an increase in the percentage of women using male condoms and a very slight increase in the percentage using an IUD, but the changes are no longer exaggerated.

If we wanted to compare the trend in male condom use with the trend in IUD use, there is one extra amendment that should be made to the graphs. The *Y* axis scales are currently different, with Figure 3.20 having a longer scale (0–30) than Figure 3.21 (0–10).

49

When comparing two or more line graphs, the X and Y axis scales should be identical. An alternative is to put all the lines on one graph, as in Figure 3.22, which shows changes in the percentages of women using five different methods of contraception. It is generally better to use colour to distinguish the different lines from each other, but, if this is not possible, different marker styles or line styles can be used on a black and white graph. If the lines are all on top of each other, you may need to adjust the scale or use separate graphs.

Figure 3.22 shows that the pill and male condoms were the most common methods used over most of the period. This was followed by male/female sterilisation, which is less and less common. The percentages for most methods fluctuate year on year.

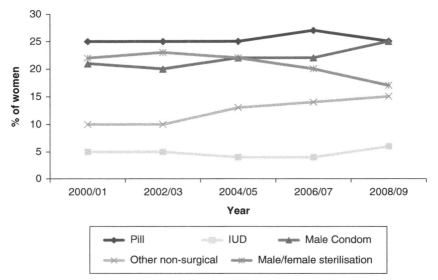

Figure 3.22 Percentage of women in Great Britain aged 16–49 using five methods of contraception, 2000/01–2008/09

Source: Lader, 2009, using figures from the Opinions Survey

Time-to-an-event data

Often in the social sciences we have data which measure the length of time that occurs before a particular event takes place. Many interesting research questions use data like these. Some examples are:

- How soon after their first birth do women have a second child?
- How quickly do workers find a new job after being made redundant?
- At what age do young people have their first sexual experiences?
- How long do children in foster care stay in one family before leaving or being moved?

Note that with such data time is measured from a starting date to the date of an event. The starting date may be obvious in some cases, for example date of redundancy or date of entering a foster family, but in other cases there is no clear starting date and a fixed time such as birth or age 16 is used as the start point.

We are usually interested in the **cumulative proportion** of people who have experienced the event by each age or after each number of months or years. To present such data graphically we can use a cumulative frequency histogram or cumulative frequency polygon. Both graph types use the same type of data and can use either frequencies or percentages.

EXAMPLE: TIME TO FIRST SNOG

Suppose a psychologist collected data on young British men and asked them at what age they had their first snog (defined as consensual kissing with a sexual meaning). Their responses, ranging from age 7 to age 20, are shown in Table 3.14. The data could be graphed as cumulative frequencies, but in this case they have been turned into cumulative percentages.

Table 3.14 Age at first snog among 42 young British men[1]: frequencies, cumulative frequencies and cumulative percentages

Age at first snog	Frequency (number of men)	Cumulative frequency	Cumulative %
7	1	1	2.38
8	0	1	2.38
9	1	2	4.76
10	0	2	4.76
11	3	5	11.90
12	6	11	26.19
13	5	16	38.10
14	6	22	52.38
15	12	34	80.95
16	2	36	85.71
17	4	40	95.24
18	1	41	97.62
19	0	41	97.62
20	1	42	100.00
21	0	42	100.00
Total	42		

[1]Sample consists of 42 men with multiple sexual partners in the previous 12 months.

Source: Hypothetical data

(Continued)

(Continued)

A cumulative frequency histogram plots the cumulative frequencies or percentages using bars in the same way as an ordinary histogram. The histogram should always go upwards from left to right, as in Figure 3.23. The graph shows that the percentage having had a first snog increases steadily with age and there is a particularly large jump between ages 14 and 15. It can be seen that 50% of the men had experienced their first snog by age 14.

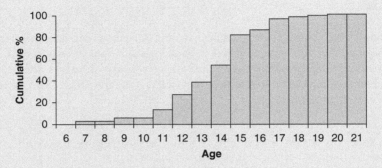

Figure 3.23   Cumulative frequency histogram of age at first snog among 42 young British men

A cumulative frequency polygon plots the same points as the histogram, but joins them with a line rather than using bars. Figure 3.24 shows the data presented in this way. Polygons tend to be used more frequently than histograms because they are clearer and more than one line can be put on the same graph. In Figure 3.25, the age at first snog is compared with the age at first sexual intercourse for the hypothetical sample of young men. As we would expect, first intercourse tends to occur at an older age than first snog. The percentage having had intercourse is very low up until age 13, after which it rises steeply and consistently to age 20 when the entire sample have experienced intercourse.

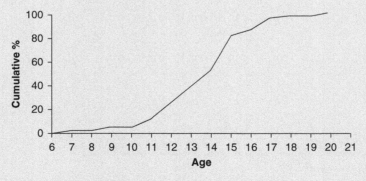

Figure 3.24   Cumulative frequency polygon of age at first snog among 42 young British men

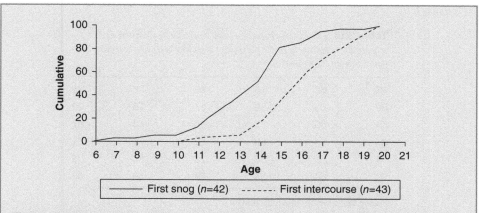

Figure 3.25 Cumulative frequency polygon showing age at first snog and first intercourse among 43 young British men

Note that in this example, all respondents experienced the event, so at the greatest age the cumulative percentage reached 100%. This is not always the case.

Summary

This chapter has introduced you to different forms of graphs and charts and when they should be used. You should now understand that certain types of variables cannot (or should not) be presented in particular ways. Many students get confused between bar charts and histograms. Remember that a bar chart is for presenting discrete data and a histogram is for continuous data. So a nominal or ordinal variable should not be presented as a histogram. Without understanding the types of data you are using it is impossible to produce effective graphs and charts. This chapter has shown you how to do this. There is more to the process than you might have first thought! As with tables, there are a number of conventions which need to be taken into account in order to present the data meaningfully. All graphs should have a clear title, X and Y axis labels, a key (legend) if necessary and the source of data if known. Always check that the graph is clear and conveys the relevant information to the reader. For instance, a pie chart without the figures is a nightmare to interpret! Now you know about the presentation of different forms of data, we can move onto some more complex ways of working with data.

PRACTICE QUESTIONS

3.1 Table 3.15 shows the percentage of 1-year-olds fully immunised against measles in the 20% of people with the lowest wealth in 36 African countries in 2010.

(a) Construct an abridged frequency table of these data.
(b) Present the data from the frequency table on an appropriate graph and comment on the distribution of the data.

(Continued)

(Continued)

Table 3.15 Percentage of 1-year-olds fully immunised against measles in African countries in 2010 among the lowest wealth quintile

73	62	28	97	26	18
89	89	39	57	78	22
57	63	73	30	54	91
65	54	89	53	75	31
9	5	86	73	65	5
92	64	84	78	58	43

Source: World Health Organization, 2012

3.2 Table 3.16 gives data for the percentage of males and females who smoke in 22 industrialised countries (2009 figures). Construct a back-to-back stem and leaf plot of the data. What does it show you about gender differences in smoking in industrialised countries?

Table 3.16 Percentage of men and women who smoke in 22 industrialised countries, 2009

	% adults who smoke	
Country	**Males**	**Females**
Canada	24	17
France	36	27
Norway	31	28
USA	33	25
Iceland	27	21
Netherlands	31	26
Japan	42	12
Finland	28	22
New Zealand	27	24
Czech Republic	43	31
Spain	36	27
Belgium	30	22
Australia	20	19
UK	25	23
Switzerland	31	21
Denmark	30	28
Greece	63	41
Italy	33	19
Portugal	32	16
Hungary	43	33
Poland	36	25
Romania	46	24

Source: World Health Organization, 2012

3.3 Present the datasets in (a) Table 3.17, (b) Table 3.18 and (c) Table 3.19 using an appropriate graphical method. You could try this by hand or using a computer package.

Table 3.17   Percentage of women's employment which is part-time, in four EU countries, 2011

Country	% of total employment
Bulgaria	2.6
Italy	29.3
Norway	42.8
UK	43.1

Source: Eurostat, 2011

Table 3.18   Acquisition of citizenship in the UK and Spain, 2003–11 (thousands)

Year	Country	
	UK	Spain
2003	130.54	26.52
2004	148.28	38.22
2005	161.76	42.86
2006	154.02	62.38
2007	164.54	71.94
2008	129.26	84.17
2009	203.63	79.59
2010	194.82	123.72
2011	177.57	114.60

Source: Eurostat, 2013

Table 3.19   Known first destinations of school leavers by region to higher education and Russell Group universities, 2009–10

Region	For students who took A levels or equivalent qualifications	
	Proportion to any higher education institution %	Proportion to a Russell Group university (including Oxbridge) %
Greater London	61	12
North West	56	10
North East	55	9
West Midlands	53	9
East Midlands	53	9
Yorkshire and Humber	53	9
East of England	50	8
South East	46	8
South West	41	7

Source: Department for Education, 2012

FOUR

AVERAGES AND PERCENTILES

Introduction

In this and the following chapter we will look at ways of describing the characteristics of a set of continuous data. We have already started to investigate the distribution of data by drawing graphs such as histograms; now we will learn how to describe the distribution of a set of data statistically, so that we can *summarise* the features of a distribution. These are known as descriptive statistics. One initial approach with descriptive statistics is to explore averages, which are also called measures of **central tendency**, to attempt to summarise the centre of the distribution of 'answers'. For instance, the mean is a commonly used measure of central tendency or average to investigate questions such as the 'average' income or 'average' waiting time for an operation. These are concepts which most people understand. But what exactly do they tell us, and are there more effective ways of describing data? For instance, is it better to talk about the most common (or modal) group when discussing average waiting times? These are all questions which need to be considered. It is also useful to think about how a distribution can be divided up into different parts so that we can say something, perhaps, about the top 5% of the distribution or the bottom 10%. For example, in the UK, the poorest 10% of households have a gross income of £216 per week or lower (when household incomes are measured net of direct taxes and inclusive of state benefits and tax credits) (Cribb et al., 2012). This is called working with percentiles.

Using these different measures enables us to condense a huge amount of data into just one or two numbers which can tell us something about the distribution. This chapter will introduce you to the process of working with descriptive statistics and averages or measures of central tendency to tell us more about the make-up or distribution of the data. It will also show you how to construct and interpret percentiles. By the end of the chapter you should be able to:

- Understand what the different measures of central tendency or averages are and how to calculate them, including the mean, median and mode
- Consider graphically how to present data with different distributions
- Calculate percentiles and understand how to present them

Measures of central tendency

When you collect data at the interval or ratio level, you will not typically get the same 'answer' each time you measure the variable. If you wanted to record how many times in a week child abuse cases were reported in the UK, the answers will vary from week to week. They might be similar, but they will certainly not be exactly the same every week. There will always be variation in reported child abuse cases from week to week and the 'answers' or figures will be distributed across a range of scores. Measures of central tendency or averages attempt to summarise the centre of these distributions of 'answers'. The overall aim is to produce a figure which best represents a midpoint in the data. So, for instance, you may want to look at finding a figure which is based on the 'average' of a year's weekly statistics about child abuse.

Other examples of when measures of central tendency are useful are for answering questions such as:

- What would a student sharing a house with friends expect to pay in rent each week?
- What do most social workers earn?
- At what age do people usually get married for the first time?

There are a number of different kinds of central tendency or averages, and we need to choose an appropriate one(s) to use: the **mean**, the **median** or the **mode**. We will use all three to answer the first question.

Table 4.1 shows the rents paid by a group of 15 second-year students living in shared houses. Of course, in practice we would like a larger sample to answer the question, but the sample has been kept deliberately small to illustrate the methods.

Table 4.1 Weekly rent paid by 15 second-year students living in shared houses, 2013 (£)

71	69	68	83	70
74	80	67	73	78
60	75	75	80	75

Source: Authors' data

The mean

The mean is the most familiar of all of the measures of central tendency and is what is most commonly meant when people refer to 'the average'.

There are two steps for calculating a mean. First, add together all the numbers:

$$71 + 69 + 68 + 83 + 70 + 74 + 80 + 67 + 73 + 78 + 60 + 75 + 75 + 80 + 75$$
$$= 1098$$

Secondly, divide this total by the number of observations (15 in this example):

$$1098 \div 15 = 73.20$$

The mean rent paid by the students is £73.20.

In statistical jargon, the formula for the mean is written:

$$\bar{x} = \frac{\Sigma x}{n}$$

To explain, the mean is usually called $\bar{x}$ and the observations are called x_1, x_2, x_3 and so on. The symbol Σ is the Greek capital letter 'sigma' and it means 'sum of', so Σx is the sum of all the values of x (the rents for each student). The letter n is simply the number of observations (students) in the sample. Note that sometimes y is used instead of x.

So the formula just describes what we did to calculate the mean: we added together all the values and divided by the number of observations. Don't be put off by the jargon: it is just a much quicker way of writing things than using words. (See Appendix C for a brief guide to mathematical notation and algebra.)

However, there are some weaknesses associated with the use of the mean. In particular it is disproportionately affected by extreme values (or outliers) in a distribution.

The mode

The mode is the number which comes up most frequently. In this dataset, two students pay rents of £80, but three students pay rents of £75. So the modal rent is £75.

If there are two numbers which come up the most frequently, the distribution is **bimodal**; in other words it has two modes.

The mode is not affected by outliers (unlike the mean), but it does not take into account all the other data.

The mode is the only measure of central tendency which can be used with nominal data, and it can also be used with ordinal data.

The median

Another way of finding an average is to find the middle observation in a set of data. The middle observation is known as the median. Half of the observations will lie above the median and half below.

The first step is to order the data:

60 67 68 69 70 71 73 74 75 75 75 78 80 80 83

In the example, there are 15 observations. Assuming that the data have been ordered, the middle observation will be the 8th observation, because there are seven observations either side of it:

60 67 68 69 70 71 73 74 75 75 75 78 80 80 83

8th

The 8th observation will be the median, so the median rent paid by the students is £74.

However, 15 is an odd number. It was easy to find the middle number because there were seven observations on each side of it. What about even numbers?

Suppose you find a 16th student to add to your sample. She pays a weekly rent of £55. The new ordered dataset would look like this:

55 60 67 68 69 70 71 73 74 75 75 75 78 80 80 83

↑ ↑

8th 9th

Now there are two observations in the middle! In such cases, we usually take the median to be halfway between the middle two, in this case the 8.5th observation. Add together the 8th and 9th observations and divide by 2 to get the median:

$$\text{Median} = \frac{73+74}{2} = £73.50$$

For any dataset, whether the number of observations is odd or even, the formula for the median can be calculated as:

$$\text{Median} = \frac{n+1}{2}\text{th observation}$$

EXAMPLE: AVERAGE HEIGHT OF A FOOTBALL TEAM

The heights of players in Liam's football team are shown in Table 4.2. What is the average height of the players?

Table 4.2 Heights of all 11 players in Liam's football team (cm)

186	179	187	205	185	184
175	178	177	175	180	

Source: Authors' data

The mean height is calculated:

$$\text{Mean} = \frac{186 + 179 + 187 + 205 + 185 + 184 + 175 + 178 + 177 + 175 + 180}{11}$$

$$= \frac{2011}{11} = 182.82 \text{ cm}$$

(Continued)

(Continued)

The mode will be 175 cm because there are two players of this height. This is clearly not a very good measure of the average, because 175 cm is actually the shortest height in the team!

To find the median, the data must be ordered:

175 175 177 178 179 180 184 185 186 187 205

With 11 observations, the median will be the (11 + 1) ÷ 2 = 6th observation. The 6th or middle observation in these data is 180 cm.

You may notice that the mean is larger than the median. This is probably because there is an outlier in the data, the player who is 205 cm tall, far taller than the rest of the team. This outlier increases the value of the mean, but does not affect the median. However, while this does make the median resistant to outliers, because it doesn't take into account the actual values within the distribution, it lacks the sensitivity of the mean.

Summary

$$\text{Median} = \frac{n+1}{2} \text{th observation}$$

$$\text{Mean} = \frac{\Sigma x}{n}$$

Mode = number which occurs most frequently

Averages for grouped data

So far we have only considered ungrouped data. Sometimes you will only have access to data which have been grouped and will need to calculate an average. Table 4.3 shows the ages of victims of violence in the 2011/12 Crime Survey for England and Wales. The data have been grouped into 10-year intervals and those under 25

Table 4.3 Age of adults who were victims of violence in the Crime Survey for England and Wales, 2011/12

All adults	Number who were victims of violence
25–34	6,740
35–44	7,586
45–54	7,603
55–64	7,650
65–74	6,807
Total	36,386

Source: Office for National Statistics, 2013c

and over 75 have been omitted in order to illustrate better the process of working with grouped data.

Let's use the three measures of central tendency we have identified to explore the characteristics of victims of crime in the survey.

The mode

The modal age of the victims of violence reported in the Crime Survey for England and Wales is 55–64. We say this because the largest number of victims of violence, 7,650, lies in this group.

However, '55–64' is not very specific! This is a major flaw with the mode. For this reason modes are rarely used with grouped data.

The mean

To calculate the mean we need to know what each of the ages is. For example, we need the answer to the following question:

- Question: what are the ages of the victims of violence in the interval 55–64?
- Answer: we don't know! They could be any age between 55 and 64!

Therefore we must make an *assumption*. We must assume that ages are evenly distributed within each interval. Therefore the mean age in an interval will be the *mid-point* of that interval.

What is the mid-point of the interval 55–64? This interval actually goes from 55 years and 0 days to 64 years and 364 days; in other words it goes from 55 to 64.99 recurring. For practical purposes, we can say that the top of the interval is 65. Therefore the mid-point between 55 and 65 is 60. A common mistake is to forget that the interval includes the bit above 64.0 and to use 59.5 as the mid-point.

Steps for calculating a mean for grouped data

Step 1

List the mid-points of each category (as in column C in the working below). These are given by:

$$\text{Mid-point} = \left(\frac{\text{top of interval} - \text{bottom of interval}}{2} \right) + \text{bottom of interval}$$

This is easier than it sounds! For example, the mid-point for 25–34 will be:

$$\left(\frac{35 - 25}{2} \right) + 25 = 5 + 25 = 30$$

Remember that 35 (34.99) is the top of the interval, not 34.

Step 2

For each category, multiply the mid-point by the number in that category (see column D in the working).

Step 3

Add together all the numbers you have just obtained in step 2 and divide this total by the total number of observations (36,386 in this case).

And there you have a mean!

To understand what has just been done, follow it on the working table:

A	B	C	D
Age	Number of victims	Mid-point of interval	Number × mid-point
25–34	6,740	30	202,200
35–44	7,586	40	303,440
45–54	7,603	50	380,150
55–64	7,650	60	459,000
65–75	6,807	70	476,490
Total	36,386		1,821,280

Mean = 1,821,280 ÷ 36,386 = 50.05

Therefore, the mean age of a victim of violence among 25–75-year-olds in the Crime Survey for England and Wales 2011/12 was 50. (Age is a special case, where we usually round down to the previous birthday.)

The median

To find the median of a set of grouped data, we must first work out which observation the median is as before. In this example there are 36,386 observations in total.

$$\text{Median} = \frac{n+1}{2} = \frac{36386+1}{2} = 18193.5\text{th observation}$$

The median lies halfway between the 18,193th and the 18,194th observation. But how can we find out what observations 18,193 and 18,194 are?

To do this, we need to calculate the cumulative frequency distribution, by adding the frequencies together going down the table as shown. The cumulative frequencies show which observations lie in or below each group: for example, 21,929 observations lie in the 45–54 group or below.

Observations 18,193 and 18,194 lie in the group 45–54, because they lie between 14,326 (the highest observation in the 35–44 group) and 21,929 (the highest observation in the 45–54 group). *But* whereabouts in the group do they lie?

Age group	Frequency	Cumulative frequency	
25–34	6,740	6,740	
35–44	7,586	14,326	
45–54	7,603	21,929	← median in here
55–64	7,650	29,579	
65–75	6,807	36,386	

Let's draw a diagram to help! The top of the line in Figure 4.1 shows the top (45) and bottom (55) of the age interval. The bottom of the line shows the highest observation in the interval (21,929) and the highest in the interval below (14,326). We mark 18,193.5 on the bottom of the line and want to work out what age it corresponds to on the top of the line.

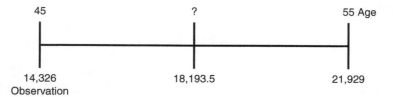

Figure 4.1 Line diagram showing the median and the cumulative frequency results of adults aged 45–55 who were victims of violence in the Crime Survey for England and Wales, 2011/12

Source: Office for National Statistics, 2013c

Now there are three distances to mark on the diagram. The distance between 45 and 55 on the top of the line is 55 – 45 = 10 (this is the width of the interval). The total distance between 14,326 and 21,929 on the bottom of the line is 21,929 – 14,326 = 7,603 (this is the frequency in the class). Finally, the distance between the

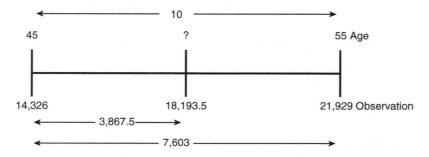

Figure 4.2 Line diagram showing the median and the cumulative frequency results including the difference between the largest and smallest figures of adults aged 45–55 who were victims of violence in the Crime Survey for England and Wales, 2011/12

Source: Office for National Statistics, 2013c

lowest observation 14,326 and the median 18,193.5 is worked out as 18,193.5 − 14,326 = 3,867.5. These distances are shown in Figure 4.2. Note that you need to divide the full distance up into 7,604 parts (one more than the frequency) so as to spread the data uniformly along the continuum.

Therefore we can say that the median is 3,867.5 ÷ 7,604 of the way along the interval:

$$\frac{3867.5}{7604} = 0.51$$

The width of the interval is 10, that is 55 − 45, so we want to move 0.51 of the way along the 10:

$$0.51 \times 10 = 5.1$$

Now add this to 45, the bottom of the interval:

$$45 + 5.1 = 50.1$$

And there we have it! The median age of a victim of violence among 25–75-year-olds in the Crime Survey for England and Wales 2011/12 was 50.1 years. (Always remember to put a proper conclusion like this at the end of your workings.)

The equation we have just calculated in stages was:

$$\text{Median} = 45 + \left(\frac{3867.5}{7604} \times 10 \right) = 50.1$$

This could be turned into an equation we could use to find any median:

$$\text{Median} = \text{Bottom of interval} + \left(\frac{\text{Distance from bottom of interval to median}}{\text{Total number of observations in interval} + 1} \times \text{Position of median width of interval} \right)$$

(Remember to divide and multiply before doing the addition.)

Using the different averages

Here are two examples to set you thinking.

EXAMPLE: LEGS

Table 4.4 gives some hypothetical data for the percentage of people in a country who have no legs, only one leg or two legs. What is the 'average' number of legs of people in this country?

Table 4.4 Number of legs of people in a country

Number of legs	% of people
0	0.01
1	0.05
2	99.94

The mean and the median of these data are as follows:

Mean = 1.99

Median = 2.00

The mean implies that 99.94% of the population have more than the 'average' number of legs!

In this case, the median, two legs, is a more sensible average, because nearly everybody has two legs.

EXAMPLE: DANCING DUDES

Like a lot of chaps I (Liam) am not the biggest fan of dancing at parties, especially without a pint or two. Most of my friends are the same. I decided to count how many times my friends danced at two separate parties. One was an 80s party (playing mainly music from the 1980s) and the other was a 90s party (playing mainly music from the 1990s) (Figure 4.3). Table 4.5 shows how the 'dancing' data looked. The mean and median number of times my five friends danced to the different types of music are calculated in Table 4.6.

Figure 4.3 Dancing dudes, 1980s and 1990s style (artwork by Alex Kanaris-Sotiriou)

Table 4.5 Number of times five friends danced to music from the 1980s and 1990s

	Friend number				
Era	1	2	3	4	5
1980s	5	7	6	55	9
1990s	6	4	10	10	11

Source: Authors' data

Table 4.6 Average numbers of times five friends danced to music from the 1980s and 1990s

	Average	
Era	Median	Mean
1980s	6.5	16.4
1990s	8.0	8.2

Source: Authors' data

Points to ponder:

- What do these results suggest about the five friends' preferences for dancing to music from the 1980s and 1990s?
- Why are the mean and median so different? Which is a better measure in this situation?

The shape of a distribution

If data are distributed in a symmetrical pattern, the mean and median will be about the same. If data are skewed, the values of the mean and median will be different because the mean is affected by the extreme values.

A histogram lets us see whether a dataset is normally distributed or skewed. The shape of a dataset can also be drawn as a sketch graph, as shown in Figure 4.4.

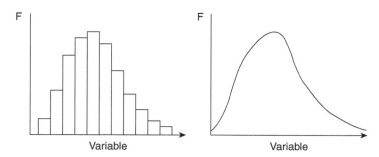

Figure 4.4 A histogram and sketch graph showing the distribution of a dataset

If data are distributed in a 'bell-shaped' pattern, as in Figure 4.5, we say that they follow a normal distribution. The normal distribution is the most important statistical distribution. Most of the observations are concentrated around the middle, with some values on either side. Data on heights or weights usually follow a normal distribution, because most people are around the average height, for example, but there are a few particularly short and particularly tall individuals in the population. When the data are symmetrical like this, the mean and median are equal.

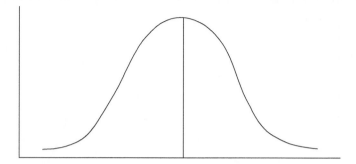

Figure 4.5 Dataset following a normal distribution

If a dataset is positively skewed, as in Figure 4.6, the data are concentrated at the lower end of the range. There are more observations where the variable takes a low value. The age of higher education students is a positively skewed variable because most are concentrated in the 18–24 age range but there are some mature students who are older than this.

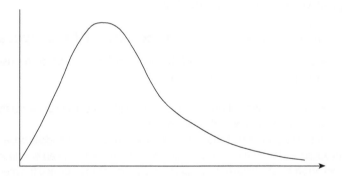

Figure 4.6 A positively skewed dataset

Where data are positively skewed, the mean will be greater than the median. The value of the mean is pulled upwards by the few very high values, whereas these do not affect the median.

For example, the median number of reported female sexual partners ever among British men in the Health Survey for England, 2012, would be considerably lower than the mean of 9.3 (Robinson et al., 2011). The mean is affected by more than 27% of men reporting 10 or more sexual partners and a small number of these men with a very large number of partners.

If a dataset is negatively skewed, the data are concentrated at the higher end of the range – in other words, there are more observations where the variable takes a high value. For example, the age at death of adults would be negatively skewed because more adults die at older ages than at younger ages. Figure 4.7 shows the shape of a negatively skewed distribution.

Where data are negatively skewed, the mean will be lower than the median. The value of the mean is pulled downwards by the few very small observations.

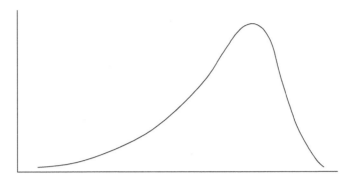

Figure 4.7   A negatively skewed dataset

Handy hint

With positively and negatively skewed distributions, it can sometimes be difficult to remember which is which. If you look at the diagrams sideways (rotate clockwise!), the positive skew looks like the letter P!

Guidelines for choosing a measure of the average

Here are some points to bear in mind when deciding whether to use a mean, a median or a mode in a particular situation:

- Modes are not used very often, though they can be useful in certain circumstances. Avoid them in general or use them along with other measures!
- The median is more intelligible to the general public because it is the 'middle' observation.
- The mean uses all the data, but the median does not. Therefore the mean is 'influenced' more by unusual or extreme data. If the data are particularly subject to error, use the median.
- The mean is more useful when the distribution is symmetrical or normal. The median is more useful when the distribution is positively or negatively skewed, because outliers do not affect the median.
- The mean is best for minimising **sampling variability**. If we take repeated samples from a population, each sample will give us a slightly different mean and median. However, the means will vary less than the medians. (For example, if we took 10 different samples of 20 students and calculated the average weight for each of the 10 groups, the mean weights would differ less between the 10 groups than the median weights.)

Percentiles

In the previous chapter we looked at cumulative percentages. These can be taken a step further with the idea of **percentiles**. What is a percentile? Percentiles divide a set of data into hundredths (100 equal parts).

> A percentile is the value at or below which a specified percentage of the scores in the distribution fall.

As an example, let's think about the 60th percentile. Here, 60% of the observations in a distribution will lie at or below the 60th percentile. Conversely, 40% of the observations will lie at or above the 60th percentile. So the 60th percentile is like the dividing value between the bottom 60% and the top 40%.

Suppose we had a set of exam marks and we knew that the 30th percentile was a mark of 52. This means that 30% of people who sat the exam got a mark of 52 or lower.

Questions: Percentiles and Income Distributions

One common use of percentiles is for describing income distributions. Table 4.7 shows some percentiles for total household income before tax and before housing costs, 2010–11, calculated by Cribb et al. (2012) at the Institute of Fiscal Studies. The 10th percentile is £216. This means that the bottom 10% of households have a weekly income of £216 or less. However, the 90th percentile tells us that 90% of households have a weekly income of £849 or less, and so from this we can work out that the top 10% of households have a weekly income of £849 or more.

Check that you can answer these questions. The answers are given after the table (no peeping!).

(a) What percentage of households have a weekly income of £319 or less?
(b) What percentage of households have a weekly income of £482 or more?
(c) What is the cut-off income for the bottom 70% of households?
(d) What is the median weekly household income?

Table 4.7 Percentiles for total household level income[1] after tax and before housing costs, 2010–11

Percentile	Weekly household income (£)
10th	216
20th	272
30th	319
40th	367
50th	420
60th	482
70th	553
80th	657
90th	849

[1]Incomes are measured net of direct taxes and inclusive of state benefits and tax credits, and at the household level.

Source: Cribb et al., 2012

69

Answers: Percentiles and Income Distributions

(a) 30% of households.

(b) 40% of households (60% receive £482 or less).

(c) 70% of households receive less than £553 per week.

(d) The median or middle observation is the 50th percentile. Therefore the median weekly household income is £420.

You may have had to think about the last question. Remember that the median is the middle of the distribution. The median is therefore the same as the 50th percentile.

Percentiles (or **centiles**) come up in many other contexts. You may have seen centile charts showing the heights and weights of babies. These enable health professionals to see whether a baby is particularly large or small.

Table 4.8 shows how percentiles have been used to describe the distribution of housing and savings assets of those aged 60–64 in the English Longitudinal Survey, 2004–5, by Banks et al. (2008) at the Financial Services Authority. It shows considerable differences in the housing/savings assets of those at the 25th percentile and the 75th percentile. This is particularly evident in relation to savings, where those at the 25th percentile have only £3,077 compared with £40,000 for those at the 75th percentile. This indicates considerable inequality in savings and has significant implications for older age, especially for the funding of care and other services.

Table 4.8 Wealth distribution and housing/savings assets of those aged 60–64 in England, percentiles in £, 2004–05

Percentile	Housing (£)	Saving (£)
25th	84,000	3,077
50th (median)	160,000	14,750
75th	250,000	40,000

Source: Banks et al., 2008

Calculating percentiles

How can we find out ourselves what the percentiles of a distribution are?

Ungrouped data

With ungrouped data, the formula is as follows:

$$\text{Value of a percentile} = \frac{\text{percentile}}{100} \times (n+1)\text{th observation}$$

where n is the total number of observations.

For example, the 30th percentile in a distribution with 200 observations will be:

$$\frac{30}{100} \times (200 + 1) = 60.3 \text{rd observation}$$

In other words, the 30th percentile will lie 0.3 of the way between the 60th and 61st observations.

You will not often need to calculate percentiles like this, and it is rather meaningless to calculate them for a very small dataset.

Grouped data

If the data are grouped, all we need to do is to calculate the cumulative frequencies and then see which group the different percentiles will lie in.

Suppose you are given the data in Table 4.9, a frequency table referring to female life expectancy at birth in 99 countries in 2009 from the World Health Organization (2012). In the third column, the frequencies have been turned into cumulative frequencies.

Table 4.9 Female life expectancy in 99 countries of the world, 2009

Female life expectancy	Frequency	Cumulative frequency
46–49	2	2
50–53	8	10
54–57	6	16
58–61	2	18
62–65	10	28
66–69	4	32
70–73	14	46
74–77	22	68
78–81	16	84
82–85	15	99
Total	99	

Source: World Health Organization, 2012

Suppose we want to know roughly where the 40th percentile lies. We can use the previous formula to find out which observation is the 40th:

$$40\text{th percentile} = \frac{40}{100} \times (99 + 1) = 40\text{th observation}$$

Which group will the 40th observation lie in? There are 32 observations in the group 66–69 or below, while the lowest 46 observations lie in the group 70–73 or below. Therefore the 40th observation must lie in the 70–73 group. So the 40th percentile lies in the group with a life expectancy of between 70 and 73 years.

We could calculate the exact values of any percentile by assuming that the observations are evenly spread within each group and interpolating as for the median. You will probably not need to do this except for the 25th, 50th and 75th percentiles, which are dealt with separately. The important thing is to understand what percentiles mean when you read statements or tables which contain them.

Special percentiles

We have already noted that the 50th percentile is the same as the median. The median or 50th percentile is probably the most used percentile as it gives an estimate of the middle of a distribution.

You may also come across deciles (which divide a distribution into ten equal parts), quintiles (which divide a distribution into five equal parts) and quartiles (which divide a distribution into four equal parts).

The most common percentiles are:

25th percentile = **lower quartile**

50th percentile = **median**

75th percentile = **upper quartile**

Figure 4.8 shows how the upper quartile, median and lower quartile divide a distribution into four equal parts.

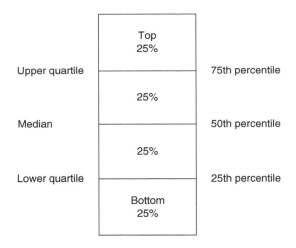

Figure 4.8   **The quartiles of a distribution**

The lower quartile

The lower quartile is the 25th percentile, so 25% of observations will lie at or below the lower quartile (LQ) and 75% at or above it. It is sometimes called the first quartile.

The formula for the lower quartile is:

$$LQ = \frac{n+1}{4} \text{th observation} = 0.25(n+1)\text{th observation}$$

where n is the total number of observations. This gives exactly the same result as the formula for the 25th percentile as follows, so either can be used:

$$LQ = 25\text{th percentile} = \frac{25}{100}(n+1)\text{th observation}$$

Let's go back to the data from the beginning of the chapter about the weekly rent paid by 16 students (the original set in Table 4.1, plus the extra student paying £55). The ordered data were as follows:

55 60 67 68 69 70 71 73 74 75 75 75 78 80 80 83

The lower quartile will be:

$$LQ = \frac{16+1}{4} = 4.25\text{th observation}$$

55 60 67 68 69 70 71 73 74 75 75 75 78 80 80 83

     ↑ ↑

    4th 5th

The lower quartile will be 0.25 (one-quarter) of the way between the 4th observation and the 5th observation. In this case, the 4th and 5th observations are £68 and £69. One-quarter of the way between £68 and £69 represents £68.25. Therefore, the lower quartile for rents is £68.25. This means that 25% of students are paying £68.25 per week or less in rent.

The upper quartile

The upper quartile (UQ) is the 75th percentile, so 75% of observations will lie at or below it and 25% at or above it. It is sometimes referred to as the third quartile.

The formula for the upper quartile is:

$$UQ = \frac{3(n+1)}{4}\text{th observation} = 0.75(n+1)\text{th observation}$$

Remember that when an equation uses brackets, this part of it must be done first.

Again, it does not matter which formula you use as they give the same result. Using the same data, the upper quartile for rents will be:

$$UQ = \frac{3\times(16+1)}{4} = 12.75\text{th observation}$$

55 60 67 68 69 70 71 73 74 75 75 75 78 80 80 83

    12th 13th

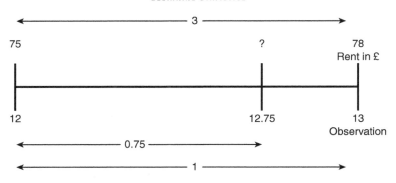

Figure 4.9 Interpolating to find the upper quartile

The upper quartile is 0.75 (three-quarters) of the way between 75 and 78. You may be able to work out in your head that the result is 77.25. If not, the way to work it out is to interpolate, as we did when calculating the median. Figure 4.9 shows how.

$$UQ = 75 + \left(\frac{0.75}{1} \times 3 \right) = 75 + 2.25 = 77.25$$

The upper quartile is £77.25, which means that 75% of the students pay £77.25 per week or less in rent.

Summary

This chapter has introduced you to ways in which we can summarise features of a distribution using descriptive statistics. In particular, it has taken you through the advantages and disadvantages of different measures of central tendency or averages, including the mean, median and mode, and when to use them. By using examples from a variety of sources it has also shown how percentiles can be created and used to tell us more about a specific part of the sample we are working with. We will now move on to look a little more about how data are spread out or distributed.

PRACTICE QUESTIONS

4.1 Table 4.10 shows Scottish Third Division football average attendances in 2012–13.

(a) Calculate the mean and the median Scottish football attendances in the Third Division.
(b) There is an outlier in the data. Which football club is this?
(c) Remove the outlier from the data and calculate the mean and median average Third Division football attendances again. Which measure of central tendency changes more with the removal of the outlier?

Table 4.10 Scottish Third Division average football attendances, 2012–13

Club	Average attendance
Annan Athletic	640
Berwick Rangers	917
Clyde	1,346
East Stirlingshire	611
Elgin City	1,030
Montrose	830
Peterhead	937
Queen's Park	2,802
Rangers	45,744
Stirling Albion	890

Source: Opta Sports Data Ltd, 2013

4.2 The data in Table 4.11 show levels of confidence in Scottish independence by age in Scotland. Calculate the mean age and median age of Scottish people who were feeling confident in Scottish independence in 2011.

Table 4.11 Confidence in independence by age, Scotland, 2011

Age group	% feel confident about independence	Total
25–34	37	143
35–44	38	212
45–54	27	227
55–64	26	193

Source: Park et al., 2012

4.3 Table 4.12 gives some percentiles for the expenditure of 'two adult retired households not mainly dependent on state pensions' in the UK, by Horsfield (2011) using family spending data from the Office for National Statistics. Answer the following questions from the table.

(a) What total expenditure difference separates the bottom 20% and top 20% of households?

(b) What is the maximum total expenditure of the bottom 20% of 'two adult retired households not mainly dependent on state pensions'?

(c) What is the difference in expenditure on recreation and culture between the bottom 20% and top 20% in the sample?

(d) What is the difference in expenditure on alcoholic drinks, narcotics and tobacco between the bottom 20% and top 20% in the sample?

(Continued)

Table 4.12   Expenditure of 'two adult retired households not mainly dependent on state pensions' by gross income quintile group in the UK, 2010 (£)

	First quintile group	Second quintile group	Third quintile group	Fourth quintile group	Fifth quintile group
Food & non-alcoholic drinks	2,558.40	2,719.60	2,948.40	3,302.00	3,998.80
Alcoholic drinks, tobacco & narcotics	795.60	452.40	556.40	473.20	1,050.40
Clothing & footwear	176.80	618.80	936.00	946.40	1,352.00
Housing (net),[2] fuel & power	2,085.20	2,116.40	2,298.40	2,771.60	4,253.60
Household goods and services	785.20	1,211.60	1,721.20	2,132.00	2,022.80
Health	67.60	369.20	442.00	353.60	1,029.60
Transport	1,274.00	1,903.20	2,496.00	3,879.20	6,744.40
Communication	494.00	436.80	535.60	473.20	910.00
Recreation & culture	1,336.40	1,924.00	3,738.80	4,248.40	7,753.20
Education	31.20	10.40	10.40	78.00	52.00
Restaurants & hotels	769.60	956.80	1,539.20	2,844.40	4,752.80
Miscellaneous goods & services	915.20	1,066.00	1,648.40	2,672.80	4,212.00
Other expenditure items	1,185.60	1,565.20	2,298.00	4,253.60	4,633.20
Total expenditure	12,469.60	15,355.60	21,169.20	28,438.80	42,754.40

4.4 Table 4.13 shows data from a survey of 103 students' TV soap watching habits in a week. Calculate the median, the lower quartile and the upper quartile for this distribution.

Table 4.13 Data on soap watching among 103 students

Average number of hours watching soaps per week	Frequency
0	20
1	10
2	24
3	15
4	13
5	7
6	5
7	4
8	0
9	3
10	2
Total	103

Source: Authors' data

FIVE

SPREADS

Introduction

While measures of central tendency (averages) are useful in describing the distribution of data, there are other useful ways of exploring their distribution. Often we need to do more than just measure averages if we want to describe data more effectively. One way of doing this is to measure the dispersion of the distribution in order to find out how spread out the data are. Fortunately, there are a number of measures which can tell us more about such dispersion. These include the range, the inter-quartile range and the standard deviation. This chapter will introduce you to these various measures of dispersion and when they can be used. By the end of the chapter you should be able to:

- Calculate a range, inter-quartile range and standard deviation
- Identify when it is appropriate to use the different measures to determine how spread out data are
- Present distributions graphically including tables of means and standard deviations and box plots

Measures of spread

The two distributions in Figure 5.1 have the same mean, the same mode and the same median. However, distribution 1 is short and fat while distribution 2 is taller and thinner. In other words, the two distributions have different spreads. So in order to distinguish between them we might want to look beyond measures of central tendency (averages).

Don't worry – you haven't just stepped into a sandwich making lesson. The **spread** is another indicator of a distribution (like measures of central tendency). A measure of spread, or dispersion, summarises how variable a distribution is.

Look at Figure 5.2 – how has the age at marriage changed between the 1970s and the 2000s? We cannot tell from the hypothetical diagrams whether the mean

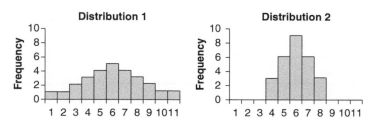

Figure 5.1 Two distributions with the same mean, mode and median

age at marriage has changed, because there are no numbers on the X axis. However, assuming that the scale is the same on both diagrams, it would appear that the age at marriage has become more variable in the 2000s, because the distribution is wider (more spread out). In other words, a larger proportion of women are marrying much earlier and much later than they were in the 1970s.

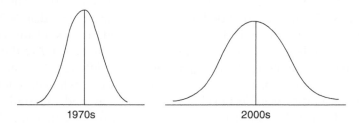

Figure 5.2 Hypothetical distribution of age at marriage for British women

Just so you know, the Office of National Statistics (2012) found that over the period 1970–2008 the mean age at marriage for both men and women generally increased, before a slight decline between 2008 and 2010. For grooms, the mean age at marriage in 1970 was 27.2 years, which compared with 36.2 years in 2010. Women have also had a similar general increase, from 24.7 years in 1970 to 33.6 years in 2010. This may be as a result of delays in first marriages and also an increasing number of remarriages (Office for National Statistics, 2012).

How can we measure how 'spread out' a distribution is? We have already used percentiles and quartiles to discover more about the shape of a distribution. But it is often useful to be able to summarise how spread out a distribution is by using just one number (rather than several percentiles or both the upper and lower quartiles).

Three possible measures which can be used are:

- the **range**
- the **inter-quartile range**
- the **standard deviation** and the **variance**

The three measures will be calculated on a dataset below in order to show how they work.

Table 5.1 Time it takes, on average, for 12 members of staff in the Department of Sociological Studies to travel to work in minutes

40	35	90	20	65	5
40	15	10	25	35	25

Source: Authors' data

As someone with a considerable commute to work, Liam thought it would be interesting to look at other members of staff in his department, Sociological Studies at the University of Sheffield, and ask how long on average it took people to travel from their home to their place of work using their most common mode of transport. Focusing specifically on social work and social policy staff, Liam asked 12 colleagues about their travel time to work (see Table 5.1). What is the average travelling time among those who responded? Do you think that the sample is representative of the travel time to work of other members of staff at the University of Sheffield?

You should be able to work out the answer to the first question. The mean travelling time is 33.75 minutes, rounded to 34 minutes, and the median is 30 minutes (check these yourself for practice). As for the representativeness of the sample, it is a rather small sample so this is problematic, as you might expect. Are there any factors which may contribute to these findings being any different than those of other members of staff? For instance, if administrative staff were included, generally lower salaries may mean they were less likely to travel as far as academic staff. In real life, it is often hard to get the data that we need to answer our research questions.

In order to get a real feel for the data we need to know how variable the travel times in the sample are. To answer this question, we will calculate the range, inter-quartile range and standard deviation for the data. In each case we will first introduce the measure.

The range

The range is calculated by subtracting the smallest value from the largest value in a sample – it's that simple!

Range = largest observation – smallest observation

For the travelling time to work data, the range will be calculated as:

Range = 90 minutes – 5 minutes = 85 minutes

The range measures the total width of the dataset. It is easy to calculate and makes intuitive sense. However, it is very sensitive to extreme values – hence it is rather susceptible to outliers. We need to treat the range with caution if these differ substantially from the rest of the distribution. For example, if a new member of academic staff joined the department and was travelling by train from London to Sheffield for work, taking on average 150 minutes, and they were added to the dataset, the range would increase to 145 minutes just because of this one new member of staff.

Inter-quartile range (IQR)

The inter-quartile range is used to overcome the main flaw of the range by eliminating the most extreme scores in the distribution. The inter-quartile range is usually used as a measure of spread if the data are skewed or the median is being used as a measure of central tendency.

The formula for the inter-quartile range (also known as the inter-quartile deviation or mid-spread) is as follows:

Inter-quartile range = upper quartile – lower quartile

To work it out we would position the variables in order from highest to lowest and concentrate on the middle 50 per cent of the distribution. So the inter-quartile range is the range of the middle half of the observations, the difference between the lower quartile and the upper quartile. The following diagram shows the idea of the inter-quartile range.

Inter-quartile range

LQ $\longleftrightarrow$ UQ

So, to find the inter-quartile range, the upper and lower quartiles must be found. This was covered in more detail in Chapter 4.

For the travelling time to work data, the 12 results must first be ranked:

5 10 15 20 25 25 35 35 40 40 65 90

Then the upper and lower quartiles can be found:

$$UQ = \frac{3}{4}(n+1) = 9.75\text{th observation}$$

The 9th observation is 40 minutes and so is the 10th observation! This makes life easy; the upper quartile or 9.75th observation will be 40 minutes.

$$LQ = \frac{n+1}{4} = 3.25\text{th observation}$$

The 3rd observation is 15 minutes and the 4th observation is 20 minutes, so the 3.25th observation will be:

LQ = 15 + (0.25 × (20 – 15)) = 16.25 minutes

Draw a diagram to help you with these calculations if necessary.

Now, the inter-quartile range (IQR) can be calculated:

IQR = UQ – LQ = 40 – 16.25 = 23.75 minutes

This means that the middle 50% of observations (travel times) have a range of 23.75 minutes.

The inter-quartile range is a useful measure of dispersion given that it is not too sensitive to outlying data values. However, as it doesn't take all of the values in the data into account, it can lack sensitivity.

The standard deviation

The standard deviation (SD) is usually used when the data are not too skewed or when the mean is being used as a measure of the average. The standard deviation tells us how widely dispersed the values in a distribution are around the mean.

The standard deviation measures the 'average' amount by which all the values deviate from the mean; in other words, it tells us something about the size of the **residuals**. A residual is the difference between a particular observation and the mean. So essentially the standard deviation is the average distance from the average. More accurately speaking, it's the square root of the averaged sum of the squared residuals! The larger the standard deviation, the greater is the spread of the data.

The standard deviation (SD) is calculated using this equation:

$$SD = \sqrt{\frac{\sum(Y_i - \bar{Y})^2}{n-1}}$$

where Y_i are the observations (the i subscript indicates *all* the values of Y rather than just one) and $\bar{Y}$ is the mean.

In words rather than jargon, the steps that must be taken to work out this formula are as follows:

Step 1

Calculate the residual $(Y - \bar{Y})$ for each observation, that is, the observation minus the mean.

Step 2

Square each residual.

Step 3

Add together all your squared residuals ($\sum$ means 'sum of').

Step 4

Divide your answer by $n - 1$, where n is the total number of observations.

Step 5

Finally, take the square root of the whole thing.

EXAMPLE: STANDARD DEVIATION OF STAFF TRAVELLING
TIME TO WORK IN MINUTES

The best way to do a calculation like this is to draw up a worksheet with a column for the observations, a column containing the mean and two columns for the first two steps of the calculation. This could be done on a spreadsheet if you prefer (if you are using a

---

spreadsheet rounding up may take place at a slightly different place resulting in minimal variation in the answer). The following shows how a worksheet would look for the calculation of the standard deviation for the travelling time data. Note that the Y_i values are the values for the 12 different staff. The mean has also been calculated beforehand.

Observation Y_i	Mean $\bar{Y}$	Residual $Y_i - \bar{Y}$	Residual squared $\left(Y_i - \bar{Y}\right)^2$
5	33.75	−28.75	826.56
10	33.75	−23.75	564.06
15	33.75	−18.75	351.56
20	33.75	−13.75	189.06
25	33.75	−8.75	76.56
25	33.75	−8.75	76.56
35	33.75	1.25	1.56
35	33.75	1.25	1.56
40	33.75	6.25	39.06
40	33.75	6.25	39.06
65	33.75	31.25	976.56
90	33.75	56.25	3,164.06
Total: $\sum(Y_i - \bar{Y})^2$			6,306.22

WORKING WITH NEGATIVE NUMBERS

You will notice that some of the numbers here are negative numbers. Most people intuitively understand what a negative number such as −5 means. Difficulties with negative numbers usually occur when a calculation is involved, for example, when two negative numbers are multiplied together, or a negative number is subtracted from another number.

The rules to follow here are:

Two signs the same → makes a positive (+)
Two different signs → makes a negative (−)

Let's first apply these rules to adding and subtracting. In this case, what matters is whether the two signs together in the middle are the same or different.

When you have two pluses together in the middle, they combine to make a plus (two signs the same). For example:

$$+8 + (+2) = +10$$

This is normally written $8 + 2 = 10$ because we usually assume that numbers are positive (+) unless we are told otherwise.

(Continued)

(Continued)

When you have a plus and a minus together in the middle, the two signs are different so they combine to make a minus. For example:

$$\text{Either } 8 + (-2) = 8 - 2 = 6$$

$$\text{Or } 8 - (+2) = 8 - 2 = 6$$

Two minuses together, however, make a plus because the signs are the same. For example:

$$8 - (-2) = 83 + 2 = 10$$

Similar rules apply to multiplying and dividing. This time the signs of the two numbers you are multiplying or dividing by are important. If both numbers are the same (both positive or both negative), the result will be positive. If one number is positive and the other negative, the result will be a negative number.
For example, multiplying:

$$3 \times 3 = 9 \text{ signs same}$$

$$-3 \times -3 = 9 \text{ signs same}$$

$$-3 \times 3 = -9 \text{ signs different}$$

$$3 \times -3 = -9 \text{ signs different}$$

For example, dividing:

$$6 \div 3 = 2 \text{ signs same}$$

$$-6 \div -3 = 2 \text{ signs same}$$

$$-6 \div 3 = -2 \text{ signs different}$$

$$6 \div -3 = -2 \text{ signs different}$$

So in our example the first residual in the table −28.75 squared is $28.75 \times 28.75 = 826.56$ squared minutes.

There are 12 observations, so $n - 1 = 11$. We now have all the numbers to put into the formula:

$$SD = \sqrt{\frac{6,306.22}{11}} = 23.94$$

Therefore, the standard deviation of the travelling times in the sample is approximately 24 minutes. We will see in a later chapter what inferences we can make as a result of knowing this.

EXAMPLE: STANDARD DEVIATION OF FOOTBALLERS' HEIGHTS IN CENTIMETRES

With practice, you will remember how to calculate a standard deviation by hand. Here is another worked example to help you get the hang of it. We will use the data on the heights of Liam's football team from Table 4.2.

Observation Y_i	Mean $\bar{Y}$	Residual $Y_i - \bar{Y}$	Residual squared $(Y_i - \bar{Y})^2$
186	182.81	3.19	10.18
179	182.81	−3.81	14.52
187	182.81	4.19	17.56
205	182.81	22.19	492.40
185	182.81	2.19	4.80
184	182.81	1.19	1.42
175	182.81	−7.81	61.00
178	182.81	−4.81	23.14
177	182.81	−5.81	33.76
175	182.81	−7.81	61.00
180	182.81	−2.81	7.90
Total: $\Sigma(Y_i - \bar{Y})^2$			727.68

$$SD = \sqrt{\frac{727.68}{11-1}} = 8.53$$

The standard deviation of the heights of Liam's football team is 8.53 cm.

Further points about the standard deviation

Working out the standard deviation by hand is time-consuming but it does help you to understand what it really means. In practice, datasets tend to be much larger and so the standard deviation is often found using a calculator or computer package. (Calculators often have two formulae for the standard deviation, so make sure you use the one dividing by $n - 1$ and not just n; using $n - 1$ gives you the standard deviation of a sample, whereas using n gives the standard deviation for a population, and in real-life data usually come from a sample.)

What does the standard deviation actually tell us? A larger standard deviation means that the data are very spread out, while a smaller standard deviation tells us that the data are quite concentrated around the mean. When comparing two distributions with similar potential ranges, the one with a larger standard deviation has a wider spread than the other.

Suppose the marks for an exam in International Relations had a standard deviation of 6.4%, while the marks for a course on Gender in Society had a standard deviation of 19.1%. This means that the marks for Gender in Society were much more variable than the marks for International Relations. More students taking Gender in Society received marks much higher or lower than the mean than students taking International Relations, where the marks were generally closer to the mean.

Another feature of the standard deviation is that it uses *all* the observations, just like the mean. Moving one observation further away from the mean will increase the standard deviation, while moving the observation closer to the mean will reduce the standard deviation.

The variance

The variance is simply the standard deviation squared. Omitting the last stage of the standard deviation formula where the value is square-rooted will give the variance instead of the standard deviation.

In the example with the travel times, the standard deviation was 23.94 minutes, but the variance found before taking the square root was 573.30 minutes.

The variance is used far less frequently than the standard deviation as a measure of spread. The reason for this is that the standard deviation is in the same units as the data and so is more easily interpretable. It is important to understand what the variance means if you see it used in a report.

The standard deviation for grouped data

In the previous chapter we calculated the mean for grouped data. What happens if you need to calculate a standard deviation for data that are grouped?

The method is similar to that for calculating the mean. Find the mid-point of each group and subtract the mean from the mid-point to obtain the residuals. Square each residual as usual and then, in an extra column, multiply it by the frequency (number in the group). Then add this column up and use it in the equation. You may wish to do this using a spreadsheet.

As a formula, this could be written:

$$SD = \sqrt{\frac{\sum[(\text{mid-point} - \text{mean})^2 \times \text{frequency}]}{n-1}}$$

Choosing between measures of spread

Range

This is not a very good measure of spread, as it only gives the extreme values, although it is easy to understand.

Standard deviation (or variance)

The standard deviation should be used with a mean as a measure of spread relative to the mean.

It makes use of all the observations, which is a good thing if the data are not skewed. However, if the data are skewed the standard deviation will be affected by outliers. Therefore it is best to use the standard deviation as a measure of spread when the data are not skewed.

The standard deviation is also preferable in terms of sampling variability. If we take lots of samples from a population, the mean value and standard deviation of each sample will be slightly different due to chance. But the means and standard deviations of the different samples will vary less than the medians and inter-quartile ranges of the different samples, so the mean and standard deviation are preferable.

Inter-quartile range

The inter-quartile range should be used as a measure of spread when the median is used as the measure of the central tendency. The inter-quartile range does not use all the observations in the dataset (it only considers the middle 50%) and so it is not affected by outliers. This means that in some respects it is more reliable as a measure of spread than the standard deviation when there are outliers or the data are skewed.

It is, however, less reliable in terms of sampling variability than the standard deviation, as explained above.

Presenting the spread of a distribution graphically

Some of the graphical methods discussed in Chapter 3 show the shape of a distribution. These include histograms and stem and leaf plots. However, there are also methods of presenting the standard deviation or inter-quartile range of a distribution using tables and box plots.

Tables of means and standard deviations

The National Fertility Studies in the USA in 1965, 1970 and 1975 collected data on the frequency of intercourse among married women. Some of the data collected are shown in Table 5.2. There appears to be a decline in the frequency of intercourse with age, but the trend over time is less clear.

Table 5.2   Mean frequency of intercourse for married women aged 30 or below in four weeks before interview, USA, 1965–75

Age	Year of Interview		
	1965	1970	1975
<19	12.5	11.6	12.1
19, 20	9.6	9.8	12.1
21, 22	9.3	8.3	10.3
23, 24	7.8	9.7	9.8
25, 26	7.6	9.4	8.9
27, 28	7.5	8.9	9.1
29, 30	6.7	8.6	8.7

Source: Trussell and Westoff, 1980

If we consider the 19–20 age group, it appears that the mean frequency of inter-course increased slightly between 1965 and 1970, from 9.6 times to 9.8 times in the four-week period. The mean frequency of intercourse then rose to 12.1 times by the 1975 survey.

But suppose we took another sample for each of the three years: would we expect to get exactly the same means? No! Each sample we take from a population will give a slightly different answer (this is known as sampling variability and we will discuss it more in Chapter 8).

The trouble here is that 9.6 and 9.8 are so close together that if different sam-ples of women had been taken we might not have seen an increase between 1965 and 1970 at all!

To get a better idea of what is really going on, we can include a measure of vari-ation with each mean. To make means more meaningful, include the standard deviation and the number in the sample for each mean in a table. The standard deviation can then tell us how variable the distribution in the sample is and the number in the sample gives an idea of how good the sample mean may be as an estimate of the population mean. Overall this can give us some idea of how much the means of different samples will vary.

Table 5.3 gives some sensible standard deviations and sample sizes for these data. The standard deviations show that there could be quite large variations in the means of different samples. So it may not be safe to say that the frequency of sex has increased between 1965 and 1970 because a slightly different sample could have led to a different conclusion.

Table 5.3 Means and standard deviations for frequency of intercourse for married women aged 19–20 in four weeks before interview, USA, 1965–75

	Year of interview		
	1965	**1970**	**1975**
Mean	9.6	9.8	12.1
Standard deviation	1.1	1.3	1.2
n	190.0	249.0	219.0

Source: Means from Trussell and Westoff, 1980; other data hypothetical

We will come back to this example in Chapter 9, where we will find out how to be confident about whether the frequency of sex increased or not.

Note that we have been assuming a normal distribution here. You may like to think about whether it is safe to do so in this instance.

Box plots

Box plots are used to present the median and inter-quartile range of a distribution, along with any outliers.

Suppose you ask 20 students how many units of alcohol they consumed in the week after exams and obtained the results in Table 5.4.

Table 5.4 Number of units of alcohol consumed by 20 hypothetical students in the week after exams

16	13	15	8	9	0	40	20	12	27
4	27	10	21	39	55	3	25	17	10

To draw a box plot, we will need to calculate the median, lower quartile and upper quartile plus some other measures.

Steps for drawing a box plot

Step 1: Order the data

0 3 4 8 9 10 10 12 13 15 16 17 20 21 25 27 27 39 40 55

Step 2: Calculate the median, lower quartile and upper quartile

Try this yourself for revision: the answers are given later on.

(a) Median =
(b) LQ =
(c) UQ =

Step 3: Calculate the inter-quartile range

(d) IQR = UQ – LQ =

Step 4: Calculate the lower and upper fences

The following are the formulae:

Lower fence = LQ – (1.5 × IQR)
Upper fence = UQ + (1.5 × IQR)
(e) Lower fence =
(f) Upper fence =

Step 5: Find the first observations inside the fences

(g) First observation above lower fence =
(h) First observation below upper fence =

Step 6: List the outliers

Outliers are observations which lie outside the two fences, in other words, observations greater than the upper fence or smaller than the lower fence.

(i) Outliers =

89

Step 7: Range

Look at the range of the data so you know what to put on the Y axis of your graph.

Step 8: Draw a box plot

Figure 5.3 shows how to plot all your answers onto a graph.

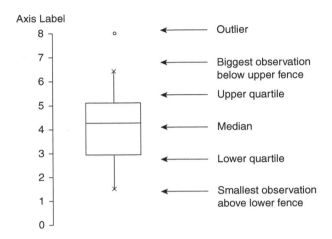

Figure 5.3   Drawing a box plot

The answers for the calculations are:

(a) Median = 15.5
(b) LQ = 9.25
(c) UQ = 26.5
(d) IQR = 17.25
(e) Lower fence = −16.625
(f) Upper fence = 52.375
(g) First observation above lower fence = 0
(h) First observation below upper fence = 40
(i)   Outlier = 55 (55 is greater than the upper fence, 52.375).

Figure 5.4 shows these results on a box plot.

Interpreting box plots

How should a box plot be interpreted? There are two important points to consider:

- Is it symmetric (in the vertical sense)? A symmetric box plot obviously represents a symmetric distribution and often implies a normal distribution. If the distribution is skewed, the box plot will not be symmetric.
- Are there outliers? Can they be explained?

The box plot in Figure 5.4 shows that the median number of units drunk in the week after exams was around 15. The box is not symmetrical; in fact it is positively skewed.

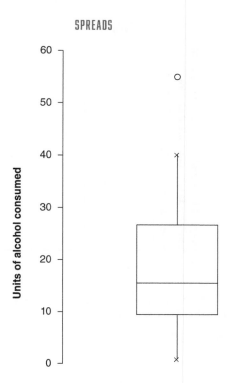

Figure 5.4   **A box plot showing units of alcohol consumed by 20 students in the week after exams**

We can tell this because the part of the box itself below the median represents a quarter of the students and the part of the box above the median represents another quarter of the students. The quarter below the median drank between 9.25 and 15.5 units in the week and the quarter above the median drank between 15.5 and 26.5 units – a wider range. Thus the quarter of students below the median are more concentrated (that part of the box is smaller) than the quarter above where the box is larger.

There is one outlier – the student who drunk 55 units during the week. Were they having a great party or drowning their sorrows?

Figure 5.5 is another example of a box plot. This has been drawn slightly differently, but can be interpreted in the same way. It shows the percentage of households which are lone-parent households with dependent children in the different London boroughs, using data from the 2011 Census.

We can see that the median percentage of lone-parent households in a London Borough is around 8%. The box plot is slightly positively skewed: all the observations below the median are concentrated between around 2% and 8%, while the observations above the median are more spread out.

Several box plots can be presented on the same graph in order to compare different groups or areas. Figure 5.6 compares the percentage of households which are lone-parent households with dependent children in London boroughs with the percentages in districts within Hampshire and Kent, using 2011 Census data. The box plots show how the distributions are in fact quite different, with much more variation between London boroughs.

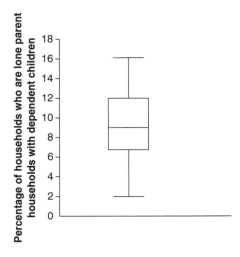

Figure 5.5 Percentage of households which are lone-parent households with dependent children in the 33 London boroughs

Source: Office for National Statistics, 2013a

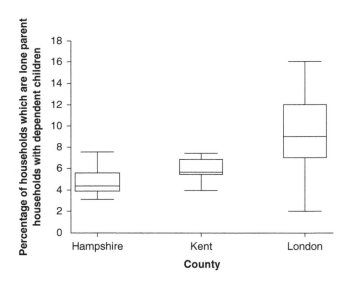

Figure 5.6 Percentage of households which are lone-parent households with dependent children in Hampshire, Kent and London

Source: Office for National Statistics, 2013a

Summary

This chapter has introduced you to the fact that we need to do more than just measure central tendency or averages if we want to describe data in a more sophisticated way. One approach is to measure the dispersion of the distribution to see how spread out the data are. This can be done in a variety of ways. These include identifying the range, the inter-quartile range, and the standard deviation. You should now know how to calculate each of these and with which kind of variables they can be used. You should also be aware of how to present distributions of spread graphically including tables of means and standard deviations and box plots. You are now ready to move on to transforming data.

PRACTICE QUESTIONS

5.1 Table 5.5 gives the marks awarded to 10 students for two pieces of assessed coursework.

(a) Calculate the mean mark for assessment 1 and assessment 2. Did the students improve between the two assessments on average?
(b) Calculate the standard deviation of the marks for each assessment. In which assessment was performance more variable?

Table 5.5 Marks awarded to 10 students in two assessments

Student number	Assessment 1	Assessment 2
1	53	55
2	61	77
3	54	57
4	59	70
5	59	61
6	48	50
7	53	41
8	49	60
9	60	17
10	60	55

Source: Authors' data

5.2 Suppose that a group of 20 students were asked the approximate number of days they smoked marijuana in the last 28 days. The hypothetical results are shown in Table 5.6.

(a) Find the median, upper quartile and lower quartile of the data and hence the inter-quartile range.
(b) Why is the inter-quartile range a better measure of spread than the standard deviation for these data?

Table 5.6 Number of days marijuana smoked in the last 28 days among 20 students

0	0	22	0	3	12	0	0	4	6
14	1	0	0	28	0	5	0	0	1

Source: Authors' data

(Continued)

(Continued)

5.3 A government target is that hospitals should treat 95% of patients arriving in accident and emergency units within 4 hours. Data are collected to determine whether this happens. Table 5.7 shows hypothetical data about waiting times in a hospital accident and emergency department. Draw a box plot of the waiting times and comment on it.

Table 5.7 Waiting times before being treated in an accident and emergency unit (hypothetical data) (minutes)

20	58	34	130	36
208	185	122	32	55
141	60	39	44	15
61	50	48	29	10
54	37	21	53	72
113	31	30	26	64

Source: Hypothetical data

SIX

TRANSFORMING DATA

Introduction

Are data sacrosanct, or can we do things with them? We can do things with them, within reason! When you look at some data you should ask yourself what are the possible errors in the data and whether the data can be 'improved'. We 'improve' data sometimes by transforming them. While not all data can be transformed, you will find out in this chapter that it can be a really useful thing to do. So how might you want to improve some data?

- You might want to make meaningful comparisons between distributions on different scales. For example, was it hotter in Miami (105°F) or Nice (29°C) on a certain day? The data could be altered to make them easier to compare, for example by converting both scales into degrees Celsius.
- You might want to highlight differences in the data rather than the data themselves, for example, to look at whether observations are above or below the mean, rather than considering the actual numbers.

Therefore, transforming variables can be a useful thing to do when dealing with statistics. These procedures do not constitute 'fiddling' the data! They just make them easier to use and understand! This chapter will introduce you to the process of transforming data. By the end of the chapter you should be able to:

- Add or subtract a constant to or from the data, or multiply or divide the data by a constant. This is known as **scaling**
- Understand the process of **standardising** the data, including being able to produce **Z-scores**

Scaling

Adding or subtracting a constant

Consider a group of women who are training for the London marathon. Their weights before they started training are shown in Table 6.1.

Table 6.1 Weights of five women before training for the London marathon (kg)

55	59	63	66	67

Source: Hypothetical data

The mean weight of the women was:

$$\text{Mean} = \frac{310}{5} = 62\,\text{kg}$$

The standard deviation of their weights is calculated as follows:

x_i	$\bar{x}$	$x_i - \bar{x}$	$(x_i - \bar{x})^2$
55	62	−7	49
59	62	−3	9
63	62	1	1
66	62	4	16
67	62	5	25
Total			100

$$\text{Standard deviation} = \sqrt{\frac{100}{5-1}} = \sqrt{25} = 5\,\text{kg}$$

So the mean weight of the women was 62 kg with a standard deviation of 5 kg.

Now suppose that the women had been running five times a week and had each lost 5 kg. Their new weights are shown in Table 6.2. We have, in effect, subtracted a constant (fixed) number (5 kg) from each weight.

Table 6.2 Weights of five women after each had lost 5 kg (kg)

50	54	58	61	62

Source: Hypothetical data

What are the mean weight and SD now?

$$\text{Mean} = \frac{285}{5} = 57\,\text{kg}$$

x_i	$\bar{x}$	$x_i - \bar{x}$	$(x_i - \bar{x})^2$
50	57	−7	49
54	57	−3	9
58	57	1	1
61	57	4	16
62	57	5	25
Total			100

$$\text{Standard deviation} = \sqrt{\frac{100}{5-1}} = \sqrt{25} = 5\,\text{kg}$$

What do you notice? The mean has decreased by 5 kg and the standard deviation has stayed the same!

Why has the standard deviation not changed? If you look at the working, you will see that the residuals $X_i - \bar{X}$ have not changed at all. Each observation is still the same distance away from the mean as before. So the shape of the distribution has not changed at all.

These facts always hold true. The rule can be summarised as follows:

Adding or subtracting by a constant:

- changes the mean by the value of the constant;
- does not change the standard deviation.

Multiplying or dividing by a constant

Suppose instead that the hard training had the effect of reducing each runner's weight by 10%. To find the new weights, the old weights must be multiplied by 0.9 to make them 10% lower. The new weights are shown in Table 6.3 and the mean and standard deviation are found as follows:

Table 6.3 Weights of five women after a 10% weight loss (kg)

49.5	53.1	56.7	59.4	60.3

Source: Hypothetical data

$$\text{Mean} = \frac{279}{5} = 55.8\,\text{kg}$$

$$\text{Standard deviation} = \sqrt{\frac{81}{4}} = 4.5\,\text{kg}$$

In fact, the mean of 55.8 kg is equal to the original mean multiplied by 0.9, as the following workings show. The standard deviation of 4.5 kg is equal to the original standard deviation multiplied by 0.9.

Mean: 62 × 0.9 = 55.8 kg

Standard deviation: 5 × 0.9 = 4.5 kg

Here you can see that a different rule applies to multiplying or dividing by a constant than that for adding or subtracting a constant:

Multiplying or dividing by a constant:

- changes the mean by an amount equal to multiplying or dividing by the constant;
- does the same to the standard deviation.

A 'multiplicative' effect like this shrinks or expands the range of the distribution. In Figure 6.1, was the constant used 1, more than 1, or less than 1?

BEFORE   AFTER

Figure 6.1   Scaling data by a constant

The constant used was less than 1, because the range has decreased. Multiplying by a constant of more than 1 will increase the range, while multiplying by a constant of less than 1 will decrease the range.

Knowing the two rules about scaling can save a lot of time. Instead of having to work out the new mean and standard deviation of a set of data that has been changed by a constant, you can simply apply the rules.

Standardising data

Sometimes you may want to compare data from more than one distribution. To do this you need to standardise the data so that they are easily comparable. This can be done by calculating **Z-scores**. A variable where the observations have been converted into Z-scores is known as a **standardised variable**.

Suppose that two friends are arguing about who did better in their optional subject exam. Claire obtained 65 in her Politics exam, while Zoe obtained 60 for Economics. Which student did better?

We could simply say that Claire did better, because she got the higher mark. But what if most students did quite well in Politics while a lot of students did badly in Economics, and so Zoe actually performed relatively well? To answer the question properly, we need to know how the other students doing those subjects performed.

On investigation, you manage to find out that for Politics the average (mean) mark was 60, with a standard deviation of 5, while for the Economics course the mean mark was 50, with a standard deviation of 6. Now we have enough information to calculate a Z-score.

A Z-score is calculated using the equation:

$$Z_i = \frac{X_i - \bar{X}}{SD}$$

where X_i is the individual mark, $\bar{X}$ is the mean mark for the whole class and SD is the standard deviation for the whole class. In words rather than jargon, the same formula can be written as:

$$Z = \frac{\text{observation} - \text{mean}}{\text{standard deviation}}$$

The Z-score for Claire will be:

$$Z = \frac{65 - 60}{5} = 1.00$$

The Z-score for Zoe will be:

$$Z = \frac{60 - 50}{6} = 1.67$$

Zoe has the higher Z-score and so she in fact did better than Claire relative to the other students in the class. The results tell us that Zoe's score is 1.67 standard deviations above the mean result for Economics, while Claire's result was only 1 standard deviation above the mean Politics mark.

In fact both the students did well, because they scored above the average and hence had a positive Z-score. Suppose a third friend, Jo, obtained a mark of 47 in the Economics exam. Jo's Z-score will be:

$$Z = \frac{47 - 50}{6} = -0.5$$

Jo had a negative Z-score, which meant that her Economics mark was below the average for the class. In fact, her mark was exactly 0.5 standard deviations below the average. This makes sense if you think about it; the standard deviation is 6, so 0.5 of a standard deviation is 3. Jo's mark of 47 is 3 marks below the mean, 50.

By now, you have probably worked out the rules for interpreting Z-scores, as highlighted in the box.

A Z-score measures the number of standard deviations an observation is away from the mean.
A positive Z-score shows that the observation is greater than the mean (above average).
A negative Z-score shows that the observation is lower than the mean (below average).
The Z-score will be zero if the observation equals the mean.

Most Z-scores will lie in the range from $Z = -2$ to $Z = 2$. Values more than two standard deviations from the mean tend to be extreme values (outliers).

Note that we are assuming here that the results for both exams were roughly *normally distributed* – in other words, a histogram of the results would be bell-shaped and not skewed. This is why it is safe to use the mean and standard deviation as measures of the average and spread.

The mean and standard deviation of a standardised variable

A standardised variable has certain properties.

Let's go back to the runners (before they were fortunate enough to lose weight!) and investigate some properties of Z-scores. The mean weight of the runners was 62 kg with a standard deviation of 5 kg. The Z-scores would be calculated like this:

X_i	$Z = \dfrac{X_i - \bar{X}}{SD}$
55	−1.4
59	−0.6
63	0.2
66	0.8
67	1.0

Try adding up all the Z-scores. What do you get?

The answer is zero! So if we wanted to calculate the mean Z-score, it would be $0 \div 5 = 0$. The mean of a set of Z-scores is always zero, so you can check that you have calculated your Z-scores correctly by seeing if they add up to zero.

To calculate the standard deviation of the Z-scores, we would have to subtract the mean (mean Z-score = 0) from each Z-score and then square it:

Z_i	$(Z_i - \bar{Z})^2$
−1.4	1.96
−0.6	0.36
0.2	0.04
0.8	0.64
1.0	1.00
	4.00

$$SD = \sqrt{\frac{(Z_i - \bar{Z})^2}{n-1}} = \sqrt{\frac{4.00}{4}} = \sqrt{1} = 1$$

Therefore the standard deviation of the Z-scores is 1.

You won't need to do this calculation, but what we have discovered is a rule which applies to any dataset (whatever its shape) once you have standardised it:

The mean of a standardised variable is 0.
The standard deviation of a standardised variable is 1.

This is useful when calculating a set of Z-scores, because you can check that your answers are right by finding their mean and standard deviation.

Z-scores are extremely useful, especially when you want to amalgamate scores on different scales (as in the first example and the example below) in order to rank, for example, people or countries in numerical order.

Calculating a Z-score index

Table 6.4 shows the distribution of marks on some introductory first-year courses taken by students in the first semester. All students take four subjects and we want to compare two students, student A and student B. Why would it not be fair just to find the average score of each student?

Table 6.4   Distribution of marks on six first-year courses

Course	Mean mark (%)	Standard deviation
Quantitative Methods	65	2
Politics	55	5
Sociology	54	4
Psychology	49	3
Economics	51	6
Demography	53	4

Source: Hypothetical data

The results would be biased in favour of those who did quantitative methods and other subjects with high mean scores. We want a score for each course which indicates the relative position of an individual on that course. Therefore we should standardise the data and calculate Z-scores.

Table 6.5 shows the actual marks obtained by the two students.

Table 6.5   Marks obtained by two students in first semester

Course	Marks for student A (%)	Marks for student B (%)
Quantitative Methods	67	–
Politics	53	52
Sociology	56	54
Psychology	43	–
Economics	–	57
Demography	–	57

Who did better overall? It is not possible to tell at a glance, so a strategy is needed! The best strategy is to calculate a Z-score for each mark and then combine them.

For example, student A got 67 for Quantitative Methods. The mean score for that course was 65 with a standard deviation of 2. The Z-score will be:

$$Z = \frac{67-65}{2} = 1.00$$

What does this tell us? It tells us that student A's mark for Quantitative Methods was exactly one standard deviation above the average mark for that course. In other words, the student did pretty well!

Try the rest yourself and see if you come up with the answers shown within the mean calculations to follow. Note that a Z-score of 0 means that a student got the average mark. A Z-score of 1 means that they were one standard deviation above the average; a Z-score of –2 means they were two standard deviations below the average; and so on.

Now we need to find the mean Z-score for each student by adding up their Z-scores and dividing by 4 (the number of exams taken).

$$\text{Student A: mean } Z\text{-score} = \frac{1.0 + (-0.4) + 0.5 + (-2.0)}{4} = -0.225$$

$$\text{Student B: mean } Z\text{-score} = \frac{-0.6 + 0.0 + 1.0 + 1.0}{4} = 0.35$$

Therefore we can conclude that student B did better than average on the chosen courses (because the mean Z-score is positive) and student A did below average when we put all the courses together (because the mean Z-score is negative). Student B performed better overall in the exams than student A.

EXAMPLE: DEPRIVATION IN AREAS OF YORKSHIRE

A common use of a Z-score index is to compare different areas. This example uses data from the 2011 Census to compare how relatively well off or deprived different areas in Yorkshire were in 2011. Social deprivation is measured here using three variables: the percentage of households who do not own a car; the percentage of residents with a limiting long-term illness; and the percentage of households not in owner-occupied accommodation. The data are shown in Table 6.6. The variables have been named NOCAR, ILL and NOTOWN for short. A high value for any of the three variables could indicate that an area is socially deprived, whereas a low value would suggest that the area is more affluent. One reason for using Z-scores is that the raw percentages on the three variables would have very different ranges. One with generally large values would swamp the index compared with one with generally small values. By standardising each one we compare like with like.

Table 6.6 Three variables measuring social deprivation in local and unitary authorities in Yorkshire in 2011

Area	Deprivation measure		
	% of households with residents with no car (NOCAR)	% of residents with limiting long-term illness (ILL)	% of households not in owner-occupied accommodation (NOTOWN)
Barnsley	26.9	23.9	35.2
Bradford	30.5	17.3	34.4
Calderdale	27.3	17.9	33.1

Area	Deprivation measure		
	% of households with residents with no car (NOCAR)	% of residents with limiting long-term illness (ILL)	% of households not in owner-occupied accommodation (NOTOWN)
Craven	17.2	17.9	26.4
Doncaster	27.8	21.6	34.3
Hambleton	13.3	16.9	30.0
Harrogate	16.4	15.5	28.5
Kirklees	26.4	17.7	32.5
Leeds	32.1	16.8	41.4
Richmondshire	13.3	15.3	35.5
Rotherham	26.6	22.0	34.5
Ryedale	14.6	17.8	32.4
Scarborough	28.8	22.5	33.7
Selby	14.9	16.4	24.3
Sheffield	33.0	18.7	41.3
Wakefield	26.9	22.0	35.9

Source: Office for National Statistics, 2013a

To find out which areas are relatively more deprived or more affluent, the three variables can be combined in a Z-score index. With a large number of observations like this, it may be better to carry out the calculations on a computer spreadsheet.

First, the mean and standard deviation for each variable must be found. These are shown in Table 6.7.

Table 6.7 Mean and standard deviation of the three variables measuring social deprivation in local and unitary authorities in Yorkshire in 2011

Area	Deprivation measure		
	% of households with residents with no car (NOCAR)	% of residents with limiting long-term illness (ILL)	% of households not in owner-occupied accommodation (NOTOWN)
Mean	23.50	18.76	33.34
Standard deviation	7.10	2.72	4.57

Source: Office for National Statistics, 2013a

Secondly, Z-scores are calculated. For example, the Z-score for NOCAR for the Sheffield district is worked out as follows:

$$Z = \frac{\text{observation} - \text{mean}}{\text{SD}} = \frac{33 - 23.50}{7.10} = 1.34$$

This is a fairly high Z-score and indicates that the Sheffield area is deprived in terms of car ownership; in other words, the percentage of people living in households with no car is quite high. However, other factors may need to be taken into account in the analysis, including remoteness in relation to amenities; for example, in areas with good public transport, more people may choose not to own a car so this may not always indicate deprivation.

The Z-scores for all three variables are shown in Table 6.8. They have been calculated to two decimal places. When calculating Z-scores, check that the values of the Z-scores look sensible – they should mainly lie between −2 and +2. Where the Z-score is particularly high or low, check the raw data and see if the district does indeed have an extremely high or low percentage of people for that variable.

The final column in Table 6.8 shows the Z-score index, where the three Z-scores for each variable have been added together and the total divided by 3 (because there are three variables). This gives us a mean Z-score or index value for each district.

For example, in the Sheffield area, the mean Z-score is calculated as follows:

$$\text{Index value} = \frac{1.34 + (-0.02) + 1.74}{3} = 1.02$$

Note that you may be thinking correctly that the mean of a set of Z-scores equals zero. This applies to the Z-scores from one distribution, but here we are combining

Table 6.8 Z-scores and Z-score index for deprivation in local and unitary authorities in Yorkshire in 2011

| Area | Z-scores | | | |
	NOCAR	ILL	NOTOWN	Deprivation index
Barnsley	0.47	1.89	0.41	0.92
Bradford	0.99	−0.54	0.23	0.23
Calderdale	0.54	−0.32	−0.05	−0.06
Craven	−0.89	−0.32	−1.52	−0.91
Doncaster	0.61	1.04	0.21	0.62
Hambleton	−1.44	−0.68	−0.73	−0.95
Harrogate	−1.00	−1.20	−1.06	−1.09
Kirklees	0.41	−0.39	−0.18	−0.05
Leeds	1.21	−0.72	1.77	0.75
Richmondshire	−1.44	−1.27	0.47	−0.75
Rotherham	0.44	1.20	0.25	0.63
Ryedale	−1.25	−0.35	−0.21	−0.60
Scarborough	0.75	1.38	0.08	0.74
Selby	−1.21	−0.87	−1.98	−1.35
Sheffield	1.34	−0.02	1.74	1.02
Wakefield	0.48	1.19	0.56	0.74

Source: Office for National Statistics, 2013a

the scores for several distributions (the three different variables) and so the mean Z-score for each district will not equal zero.

What do these values tell us? For all three variables, a high Z-score indicates deprivation, while a low score indicates relative affluence. So an area with a positive index value will be relatively deprived and a district with a negative index value relatively affluent. Sheffield is the most deprived area, with an index value of 1.02, with Leeds, Barnsley, Scarborough and Wakefield also having high index values. This may be what you expect from larger cities. However, if we investigate the findings further we may want to include other indicators, as the fact that predominantly urban Sheffield and Leeds both have significant student populations, with two or more universities, is likely to impact on the figures. For instance, owner occupation and car ownership are less common among students than most other groups. On the other hand, they both have lower than average percentages of people living with a long-term limiting illness. Looking at the negative index values, the most affluent ward appears to be Selby with a score of −1.35, followed by Harrogate with a score of −1.09. These are known as relatively affluent areas.

Further notes on creating a *Z*-score index

In the deprivation index calculated for Yorkshire districts each variable had equal weight. However, if you think that unemployment, for example, is a particularly important indicator of deprivation, then you could give it extra weight by, for example, doubling the Z-score for that variable in each area. You would then need to divide the sum of the Z-scores by 4 instead of 3 to make the index, because in effect you are using four Z-scores (two ordinary Z-scores and one which has been doubled). As an aside, weighting is also often used in sample surveys to take into account the fact that we often get different response rates among different subgroups in the population.

In this example we were also fortunate with the deprivation index calculations because, for all three variables, a high value indicated deprivation. Suppose that instead of long-term illness, the variable used was the proportion of males in employment. A high value of male employment indicates affluence rather than deprivation. To overcome this problem, after calculating the Z-scores in the usual way, the signs can simply be changed. In other words, all the positive Z-scores for that variable become negative and all the negative Z-scores become positive. A high Z-score for the variable representing male employment would now indicate deprivation rather than affluence, so all the variables are working in the same direction. For an example of this, try practice question 6.2.

Summary

This chapter has introduced you to ways in which data can be transformed to make them easier to use and understand. While not all data can be transformed, it has shown why we may want to scale or standardise the data where this is possible and how to do this. It has also shown, using deprivation indictors as an example, for instance, how to use *Z*-scores to standardise data. While a number of these processes initially look rather complex you can see that by taking things step by step, transforming data can be less complicated than first thought.

6.1 As a connoisseur of chocolate cakes, which you save to eat while doing statistics exercises, you want to investigate the prices of large chocolate cakes in the shops around the town. On your hunt, you manage to sniff out the prices of 15 brands of chocolate cake. Their prices are shown in Table 6.9.

(a) What is the mean price of a chocolate cake (to the nearest penny) and what is the standard deviation of the prices?
(b) Suppose that, due to inflation, all the shops put up their cake prices by 20p. What is the new mean price and the new standard deviation? (You should not need to do all the calculations again.)
(c) On top of this, the government decides to put VAT on chocolate cakes (crisis!), so all the prices increase by 20%. What will the mean and standard deviation be now?

Table 6.9 Prices of 15 brands of large chocolate cakes (p)

99	220	220	169	195	250	289	175
100	240	170	200	175	399	499	

Source: Hypothetical data

6.2 You have been asked by a human rights group to give some idea of how Asian countries compare in terms of female empowerment. You find comparable data on three topics: women as a percentage of the labour force; the percentage of seats in parliament held by women; and the maternal mortality rate (rate of death during childbirth). The data for eight Asian countries are shown in Table 6.10. Calculate a female empowerment index for the eight countries using *Z*-scores. In which countries are conditions most favourable for women? In which are conditions least favourable?

If you are panicking at this point, here are some hints to help you along the way!

(a) Calculate the mean score for each variable.
(b) Calculate the standard deviation for each variable.
(c) Calculate the *Z*-scores. Check that you have calculated them correctly by adding up the *Z*-scores for each variable: the total should be zero!
(d) Think about what the variables mean. If women as a percentage of the total labour force is high and the percentage of seats in parliament held by women is high, this is a good thing for women. However, if the maternal mortality rate is high, this is a bad thing for women. Therefore reverse the signs of the *Z*-scores for maternal mortality so that pluses become minuses and minuses become pluses. Now a positive *Z*-score for maternal mortality means that mortality is low. A positive *Z*-score for all three variables now indicates a good outcome for women.
(e) Find the mean *Z*-score for each country.
(f) Interpret your answers. What does a high index value or a low index value indicate?

Table 6.10 Measures of female empowerment in eight Asian countries in 2010

Country	Women as % of labour force	% seats in parliament held by women	Maternal mortality rate (per 100, 000 live births)
Philippines	38.8	22.1	99
India	25.3	10.8	200
Thailand	45.7	13.3	48
Malaysia	35.8	9.9	29
Indonesia	38.2	18.0	220
China	44.6	21.3	37
Bangladesh	39.9	18.6	240
Pakistan	20.7	22.2	260

Source: World Bank Statistics, 2012

SEVEN

THE NORMAL DISTRIBUTION

Introduction

We have already (briefly) met the normal curve. It looks rather like one of the rolling hills found in the Lake District (Figure 7.1). Graphically the normal distribution is best described by a 'bell-shaped' curve. Whether you think that mountain climbing is 'normal' behaviour or not (Julie does!), the normal (or Gaussian) distribution is extremely useful, because many variables are distributed in this way or approximately so, for instance, heights of people, blood pressure or marks in a test. The term 'normal distribution' refers to a particular way in which observations will tend to 'pile up' around a particular value rather than be spread evenly across a range of values. It is generally most applicable to continuous data. This chapter will introduce you to how to calculate whether data are normally distributed and how measures of standard deviation and Z-scores are used to do this. It will also show how we can predict the probability of an event occurring when data are normally distributed. By the end of the chapter you should be able to:

- Recognise visually whether data are likely to be normally distributed
- Calculate using the standard deviation and mean whether data are normally distributed
- Understand how and when to employ normal tables and Z-scores

Characteristics of a normal distribution

Some features of the normal curve shown in Figure 7.2 are:

- The mean lies in the middle and the curve is symmetrical about the mean. The mean is equal to the median.
- Most of the observations are close to the mean, so the frequency is high around the mean. There are fewer observations that are much greater or much smaller than the mean.
- The curve never actually touches the X axis on either side; it just gets closer and closer.

Normal curves are all the same basic shape, but can be tall and thin or short and fat. Both the curves in Figure 7.3 are normal and have the same mean, but we know

Figure 7.1 Scaling the heights of normality

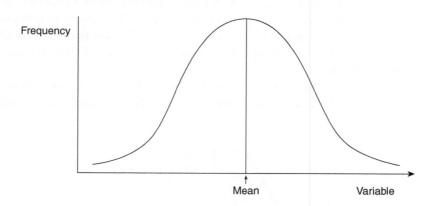

Figure 7.2 The normal curve

that the dotted curve has a larger standard deviation (greater spread) because it is more spread out and wider at the base.

A normal curve can be defined uniquely by its mean and standard deviation. Therefore if we are given the mean and SD we can construct a normal curve exactly. Because the basic shape is fixed, the proportions of the area under the curve

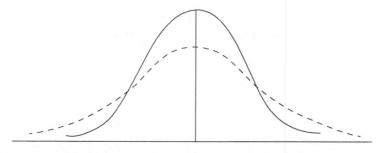

Figure 7.3 Two normal curves with the same mean but different standard deviations

falling between various points can be calculated. For example, if the variable in question was height and we knew the mean and standard deviation of heights in the population, we could work out what proportion of people were over 200 cm tall or between 160 cm and 180 cm tall. If we were interested in the leisure activities of men and women over the age of 65 and we knew the mean number of days leisure activities were undertaken in an average month and standard deviation of leisure activities in the population, we could work out what proportion of people had a hectic social life and what proportion didn't.

Areas under the normal curve

We call the total area under the curve 1.0, as in Figure 7.4, so any proportion of the curve will be a number between 0 and 1.

Half of the area lies above the mean and half below the mean, so we can say that the proportion of the area greater than the mean is 0.5. The proportion below the mean is also 0.5. This is shown in Figure 7.5. We can also say that the **probability** of an observation being greater than the mean is 0.5, because half of the observations lie above the mean.

Other areas below the curve can be described in terms of standard deviations away from the mean. These areas are the same whether a normal curve is tall and

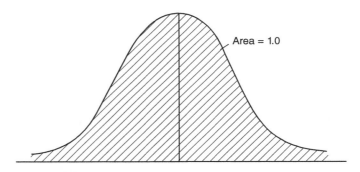

Figure 7.4   The area under a normal curve is 1

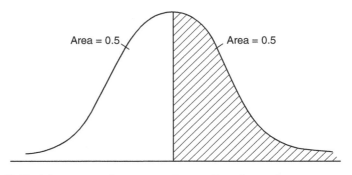

Figure 7.5   Half of the area under a normal curve lies above the mean

thin or short and fat. Some important areas are shown in Figure 7.6. This shows, for example, that:

- 0.682 of the area lies between $\bar{X}$ – SD and $\bar{X}$ + SD. This means that 68.2% of observations lie within one standard deviation either side of the mean.
- 0.954 of the area lies between $\bar{X}$ – 2SD and $\bar{X}$ + 2SD. This means that 95.4% of observations lie within two standard deviations either side of the mean.

(Remember: to convert a proportion to a percentage, simply multiply by 100.)

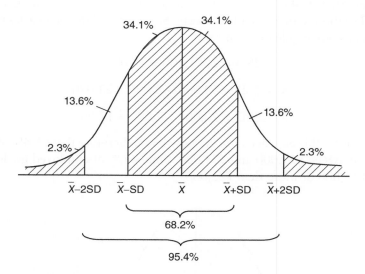

Figure 7.6   Some areas under the normal curve

Questions: Areas under the normal curve

See if you can answer these questions now. You will probably find it helpful to sketch a quick diagram each time, shading in the area that you are trying to find the size of.

(a) What proportion of the total area will lie between the mean and one standard deviation above the mean?
(b) What proportion of the total area will lie between the mean and one standard deviation to the left of the mean?
(c) What proportion of the total area will lie to the left of one SD below the mean?
(d) What proportion of the total area will lie outside the region between $\bar{X}$ – 2SD and $\bar{X}$ + 2SD?

The answers are given at the end of the chapter. These ideas may be confusing to start with, but it is all logic. Remember that the curve is symmetrical, so any area you have worked out to the right of the mean will be the same size as the opposite area on the left-hand side of the mean.

What use is all this? Let's look at two examples.

EXAMPLE: GRADUATE SALARIES

Suppose you have been told that full-time salaries for new graduates follow a normal distribution with a mean of £20,000 (the actual figure was £19,935 in 2012 according to the Higher Education Careers Services Unit (HECSU, 2012) and the Education Liaison Task Group, if you are interested!) and an SD of £1,500. What can we discover about graduate salaries by looking at the normal curve?

We know that 95.4% of observations in a normal distribution lie between two standard deviations either side of the mean. So let's calculate the salaries two standard deviations either side of the mean:

$$2SD = 2 \times 1,500 = 3,000$$

$$\bar{X} + 2SD = 20,000 + 3,000 = 23,000$$

$$\bar{X} - 2SD = 20,000 - 3,000 = 17,000$$

Figure 7.7 shows this on a diagram. We now know therefore that 95.4% of graduate salaries lie between £17,000 and £23,000. Only 4.6% of employed graduates earn less than £17,000 or more than £25,000.

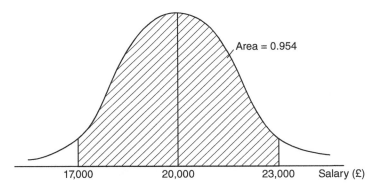

Figure 7.7 Graduate salaries

EXAMPLE: LENGTH OF STAY

The manager of a short-term nursing home for older people keeps records on how long residents stay at the nursing home before moving into long-term care or hospital or returning home. She tells you that the mean length of stay is 38 days with a standard deviation of 12 days and that the distribution of the data is normal.

We know that 68.2% of the area lies within one standard deviation.

$$\bar{X} + SD = 38 + 12 = 50 \text{ days}$$

$$\bar{X} - SD = 38 - 12 = 26 \text{ days}$$

Therefore we can say that 68.2% of the residents stay between 26 and 50 days at the home. This is shown in Figure 7.8.

What is the probability of a resident staying for more than 62 days? We have that 62 days is the mean (38) plus two standard deviations ($2 \times 12 = 24$). We know that the area above $\bar{X} + 2SD$ is 0.023. So only 2.3% of residents stay for more than 62 days.

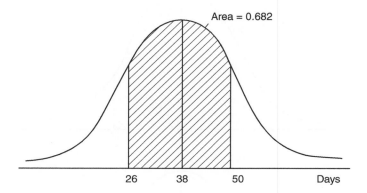

Figure 7.8 Length of stay

Z-scores and the normal distribution

We already know that when we standardise a variable by turning the values into Z-scores, the mean of the Z-scores is 0 and the SD is 1 (see Chapter 6). The distribution of a standardised variable is known as the 'standard normal distribution' and is pictured in Figure 7.9.

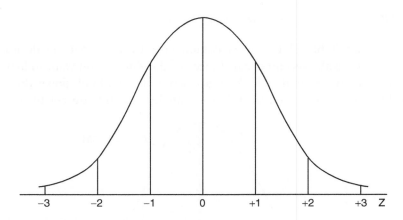

Figure 7.9 The standard normal distribution

Remember that a Z-score of +1 indicates that you are one standard deviation above the mean. So the area between the mean and Z = +1 will be 0.341; this is the same as saying the area between $\bar{X}$ and $\bar{X} + SD$ is 0.341.

Unfortunately, Z-scores do not always take nice round values like +1. To find the area between the mean and any Z-score, we must use a **normal table**. A normal table can be found in Appendix 1 (Table A1.1).

A normal table gives the proportion of the area of the curve between the mean and a certain positive value of Z, as shown in Figure 7.10. This is the same as the proportion between the mean and the same negative value of Z.

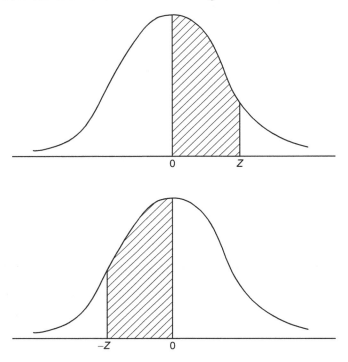

Figure 7.10   Area between the mean and a Z-score

Have a look at Table A1.1. The first column gives the Z-score to one decimal place. The top row gives the second decimal place of the Z-score you want to look up.

So suppose you wanted to look up a Z-score of 1.05. Look down the first column till you reach 1.0, then travel across the 1.0 row till you get to the column

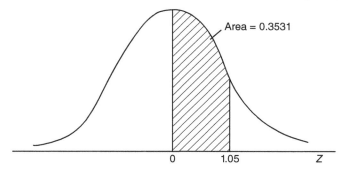

Figure 7.11   Area between the mean and a Z-score of 1.05

with 0.05 at the top. You should then be able to read off the area figure for a Z-score of 1.05, which is 0.3531.

This means that the proportion of the area that lies between the mean and $Z = 1.05$ (1.05 standard deviations above the mean) is 0.3531. This is shown in Figure 7.11. Sketching a quick diagram is always a good idea when finding an area like this.

Questions: Areas between mean and Z-score

Try these examples to check that you can do this. The answers are at the end of the chapter.

(a) What is the proportion between the mean and $Z = 2.00$?
(b) What is the proportion between the mean and $Z = 0.59$?
(c) What is the proportion between the mean and $Z = 1.66$?
(d) What is the proportion between the mean and $Z = -0.03$?

Now, can you work out how to calculate the area to the *right* of $Z = 2.00$? Again, always draw a diagram first (see Figure 7.12). Your diagram does not need to be perfect: you just need to be sure that you shade in the correct area. So:

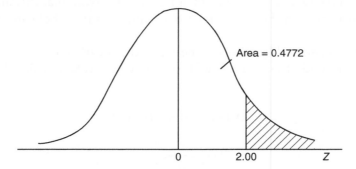

Figure 7.12   Area to the right of $Z = 2.00$

- The area between the mean and $Z = 2.00$ is 0.4772.
- The whole area to the right of the mean is 0.5 (half).
- Therefore the area to the right of $Z = 2.00$ must be $0.5 - 0.4772 = 0.0228$.

Questions: Areas outside Z-scores

Try the following questions for practice. Again, the answers are at the end of the chapter.

(a) What is the area to the right of $Z = 1.00$?
(b) What is the area to the right of $Z = 2.33$?
(c) What is the area to the left of $Z = -0.86$?
(d) What is the area to the left of $Z = -1.17$?

The following two examples show how calculating the areas below the normal curve can be useful in different situations.

EXAMPLE: QUANTITATIVE METHODS RESULTS

Suppose you get really excited doing all these practice questions and end up scoring 80% in your Quantitative Methods exam. It sounds pretty brilliant. But what if everyone else found it really exciting too and did better than you? You go off and discover that the mean mark for the class was 74 with a standard deviation of 4. You have to assume that the marks are normally distributed. What is your Z-score?

$$Z = \frac{\text{your mark} - \text{mean mark}}{\text{standard deviation}} = \frac{80-74}{4} = 1.5$$

With a Z-score of 1.5, you should be very happy! You scored 1.5 standard deviations above the average. But you are still curious to know how many of your friends you beat! What we need to know is the proportion of marks that come below Z = 1.5.

In notation, we want to know P(Z < 1.5), the probability that Z is less than 1.5. The area below Z = 1.5 is shaded in Figure 7.13. Note that this includes not only the area between the mean and Z = 1.5, but also the whole area below the mean.

From the normal tables, the area between the mean and Z = 1.5 is 0.4332. To this we must add the area below the mean, which is half of the observations or 0.5.

$$0.4332 + 0.5 = 0.9332$$

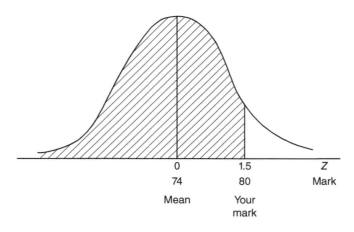

Figure 7.13 Area below Z = 1.5

This means that 93.32% of the class got a lower mark than you (less than 80). So you should go and party!

EXAMPLE: MOTORWAY DRIVING

Suppose a researcher wanted to investigate the speed of driving on motorways. She took a sample of cars driving down her local motorway one morning and measured how fast they were travelling. She found that the speed of the cars in her sample followed a normal distribution with a mean of 75 mph and a standard deviation of 8 mph. What proportion of cars were breaking the legal speed limit of 70 mph?

Draw a diagram: We want to know the proportion of cars going at over 70 mph, so we need to find the shaded area shown in Figure 7.14.

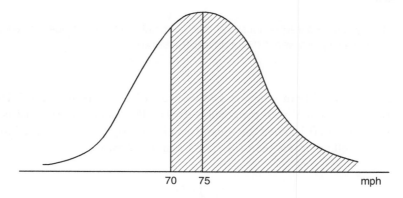

Figure 7.14 **Proportion of cars breaking the speed limit**

Calculate the Z-score:

$$Z = \frac{70 - 75}{8} = -0.625$$

Find the area: We need to find the area between the mean ($Z = 0$) and $Z = -0.625$. On Table A1.1 we can look up the values for 0.62 or 0.63, so to find the value for 0.625 we must **interpolate** between the two (as we did when calculating medians and percentiles). The value for $Z = 0.625$ will be halfway between 0.2324 and 0.2357, so we can calculate it like this:

$$\text{Area} = \frac{0.2324 + 0.2357}{2} = 0.2341$$

We must now add on 0.5 for the area to the right of the mean.

$$\text{Total area} = 0.5 + 0.2341 = 0.7341$$

Draw a conclusion. We have found that 73.41% of car drivers in the sample that morning were breaking the speed limit.

Questions: Speeding motorists

Now see if *you* can find:

(a) The proportion of motorists who were doing 90 mph or more.
(b) The proportion of cars being driven at speeds below 65 mph.

The answers are at the end of the chapter.

From your own experience, do you think that the researcher's sample was a representative sample of motorway driving speeds?

EXAMPLE: MOTORWAY DRIVING IN REVERSE!

Using the figures from the previous example, we might want to ask a question such as: at less than what speed were 60% of drivers travelling?

First of all we could put what we know onto a diagram, as in Figure 7.15. Here 60% of the area is equal to a proportion of 0.6. The shaded area consists of 0.5 below the mean and 0.1 above the mean (which adds up to 0.6). We are interested in the 60% travelling at less than the speed marked '?' mph.

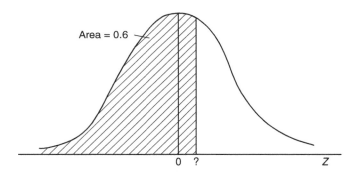

Figure 7.15   **Highest speed at which the slowest 60% of drivers were travelling**

In the previous examples, we worked out the Z-score and then looked up the area on the table. Now we must work backwards. We know the area, 0.1, and want to find the corresponding Z-score. In other words, what is the Z-score which corresponds to an area of 0.1 above the mean?

This time, look up the area 0.1 in the *body* of the normal table. The closest value to 0.1 is 0.0987, which will do as an approximation. If you then look along the row and up the column, you will see that this corresponds to a Z-score of 0.25.

Now that we have a Z-score, we can put what we know into the usual equation for Z (we know the answer, but not the value of '?').

$$Z = \frac{\text{observation} - \text{mean}}{\text{standard deviation}}$$

$$0.25 = \frac{? - 75}{8}$$

With a bit of rearranging it is now possible to find '?':

$$0.25 \times 8 = ? - 75$$

$$(0.25 \times 8) + 75 = ?$$

$$77 = ?$$

Therefore we can say that 60% of cars were travelling at a speed of less than 77 mph.

EXAMPLE: THE FASTEST DRIVERS

You want to know what speed the fastest 20% of cars are travelling above.

Figure 7.16 shows the top 20% or 0.2 of the area on a diagram. We cannot find this area from the tables; instead we must find the area between the mean and '?' which is 0.5 − 0.2 = 0.3.

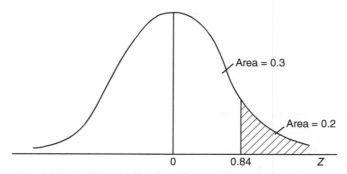

Figure 7.16   Lowest speed at which the fastest 20% of drivers were travelling

If you look up the area 0.3 in the body of the normal tables, you will find that the closest value is the area 0.2995. This corresponds to a Z-score of 0.84. To what speed does a Z-score of 0.84 correspond? To find out, everything we know must be put into the equation for Z as before:

$$0.84 = (? - 75) \div 8$$

$$0.84 \times 8 = ? - 75$$

$$(0.84 \times 8) + 75 = ?$$

$$81.72 = ?$$

Therefore the top 20% of cars were travelling at over 81.72 mph.

Question: Inter-quartile range for driving speeds

Can you think how you would calculate the inter-quartile range for driving speeds in the sample?

The IQR is the difference between the upper quartile (75th percentile) and the lower quartile (25th percentile), so you would want to find the Z-scores corresponding to the area of 0.25 above the mean and 0.25 below the mean, as in Figure 7.17. The Z-scores would then be converted to actual speeds as in the examples above, and finally the speed that represents the lower quartile subtracted from the speed which equals the upper quartile to get the IQR.

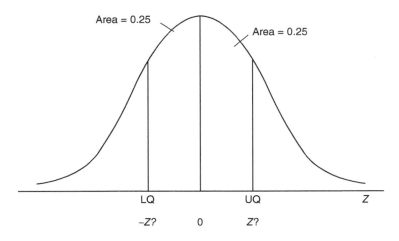

Figure 7.17   The inter-quartile range on a normal curve

Try this yourself: the answer is with the others at the end of the chapter.

Answers: Areas under the normal curve

(a) 0.341.
(b) 0.341. The curve is symmetrical, so the answers to (a) and (b) are the same.
(c) 0.159 (0.136 + 0.023).
(d) 0.046 (0.023 + 0.023).

Answers: Areas between mean and Z-score

(a) 0.4772.
(b) 0.2224.
(c) 0.4515.
(d) 0.0120. This is the same as the area between the mean and $Z = 0.03$ due to symmetry.

Answers: Areas outside z-scores

(a) 0.1587 (0.5 – 0.3413).
(b) 0.0099 (0.5 – 0.4901).
(c) 0.1949 (0.5 – 0.3501).
(d) 0.1210 (0.5 – 0.3790).

Answers: Speeding motorists

$$\text{(m)} \quad Z=\frac{90-75}{8}=-1.88$$

$$\text{Area} = 0.5 - 0.4699 = 0.0301$$

(a) Therefore 3.01% of motorists were going at more than 90 mph.

$$\text{(n)} \quad Z=\frac{65-75}{8}=-1.25$$

$$\text{Area} = 0.5 - 0.3944 = 0.1056$$

(b) Therefore 10.56% of motorists were travelling at less than 65 mph.

Answers: Inter-quartile range for driving speeds

The Z-score for the area 0.25 above the mean is 0.67 (closest). The Z-score for 0.25 below the mean will be –0.67.

$$\text{Upper quartile}: \quad 0.67=\frac{?-75}{8}, \text{ so } ? = 80.36 \text{ mph}$$

$$\text{Lower quartile}: \quad -0.67 =\frac{?-75}{8}, \text{ so } ? = 69.64 \text{ mph}$$

$$\text{Inter-quartile range} = 80.36 – 69.64 = 10.72 \text{ mph}.$$

Summary

This chapter has introduced you to the normal distribution of data and its characteristics. It is made up of data which 'piles' up around a particular value (in the shape of a nice hill). When data are distributed in this way they can be used to identify the likelihood of particular patterns or events taking place when we have information about the mean and standard deviation. The use of Z-scores, first introduced in Chapter 6, and a normal table to assist with this process have also been highlighted.

7.1 Intelligence quotients (IQs) are known to be approximately normally distributed with a mean of 100 and standard deviation of 15.

(a) What two values will the middle 68.2% of the population's IQs lie between?
(b) What proportion of the population will have an IQ below 80?
(c) What proportion of the population will have an IQ below 110?
(d) What proportion of the population will have an IQ between 95 and 115?
(e) What proportion of the population will have an IQ between 70 and 85?
(f)   What IQ value (to one decimal place) will the top 10% of the population have an IQ greater than? (Find the closest value of Z in the table.)

7.2 The time it takes students to walk from a hall of residence to the university campus is normally distributed with a mean of 19 minutes and a standard deviation of 3 minutes.

(a) What proportion of students take less than 22 minutes to reach the campus?
(b) What proportion of students take more than 15 minutes to reach the campus?
(c) What proportion of students take less than 17 minutes to reach the campus?
(d) If all students leave 20 minutes before a lecture is due to start, what proportion of them will be late?
(e) What proportion of students take between 21 and 25 minutes to reach the campus?

7.3 In a first year Sociology course, the exam marks were normally distributed with a mean of 58% and a standard deviation of 10%.

(a) Students who get less than 35% are deemed to have failed. What proportion of students failed the exam?
(b) If an A grade is awarded to all students achieving 70% or higher, what proportion of students got an A?
(c) The best 40% of students achieved better than what mark?

EIGHT

FROM SAMPLES TO POPULATIONS

Introduction

In social science we frequently want to find out something about a particular group of people. For example, we may want to find out which party people are planning to vote for the day before an election, or what proportions of adults believe that cannabis should be legalised. We could ask everybody in the country these questions. However, it would be rather impractical; the cost would be immense! The good news is that we don't need to since, properly drawn, representative samples vary in a systematic way and make good estimates of the things in the **population** we are trying to find out about. Using a sample to estimate values of a population saves vast amounts of time and money. Here are some examples of populations and samples that a researcher may choose to use:

---

1   What is the average income of full-time workers in Great Britain? Population: all full-time workers in Great Britain. Sample: 1000 full-time workers selected randomly by their National Insurance numbers.
2   How much time do teenagers in England doing GCSEs spend doing homework each day? Population: all teenagers in England doing GCSEs. Sample: a sample of 600 teenagers in England in Years 10 and 11 from a selection of different schools throughout the country.
3   What proportion of British MPs would vote for entry into the euro? Population: all British MPs. Sample: 100 MPs skilfully selected to reflect MPs in different parties and regions of Britain.

However, it is not always as easy as it sounds. Making sure your sample is representative and actually represents the population it is trying to is not straightforward. For instance, if you were interested in students' perceptions of their future job prospects and your sample was based on students doing medicine at Cambridge, their perceptions are likely to be different than those of students doing other courses (which have less of a defined career path) and possibly at other universities (not naming names!). There are also several different ways of finding a sample, which can impact on the kind of responses you receive and how likely it is to be able to generalise to the population. These will be discussed in this chapter, and vary from rather systematic to unsystematic approaches. In addition, sample means will be considered, including how these can be used to predict particular outcomes. By the end of the chapter you should be able to:

- Understand the difference between a sample and a population
- Identify **sampling errors**
- Acknowledge different forms of sampling
- Calculate sampling means and understand how these can be used to predict particular outcomes

Sampling

Identifying the sample you wish to use is a careful process and requires considerable thought to ensure clarity. Before selecting the sample, it is critical that the population is defined appropriately. Once you have decided who is included in the population, a sampling frame needs to be obtained – this is a list of everybody in the population. You need to consider all subgroups in the population. Sampling frames typically include coverage errors. For instance, under-coverage occurs when particular subgroups in the population are omitted from the sampling frame. If you were researching teenage drug use and your sample was based on those attending school, the sampling frame omits those who left school early. Drug use and other patterns of behaviour may be very different among this group. Another error is when people are double-counted in a sampling frame. For instance, if you were interested in this topic and also included people attending youth clubs in the sample and someone who went to school and the youth club was questioned at both, this would be an example of double-counting and problematic for the sampling frame. This emphasises the importance of finding out as much as possible about your sampling frame in order to identify possible coverage errors.

Let us think about the first example in the introduction. Can you think of any possible problems with our sample that need to be clarified?

Firstly, the population must be very clearly defined. For example, regarding 'full-time' workers:

- How many hours constitutes full-time?
- Is overtime included?

Secondly, the property of interest must be very clearly defined. For example, take 'average income':

- What do we mean by 'average': the mean or the median?
- What do we mean by 'income': does it include overtime or a company car, for example?

Thirdly, the sample must be representative of the population. Asking 1,000 people in Aberdeen, for example, what their average income is would not give you a sample that is representative of Great Britain as a whole.

We need to choose our sample very carefully in order to minimise:

- sampling error
- sampling variability

This jargon is explained below.

Types of error in samples

There are several reasons why the characteristics of a sample will not be identical to the characteristics of the population. Here three types of error are described.

Sampling variability

If we take repeated samples from the same population, the means and standard deviations of the different samples will not be the same.

For example, suppose we are interested in estimating the mean height of a lecture class. We might believe that we could choose one row randomly and take the mean height of people sitting in that row as an estimate. We would expect each mean would be different due to chance and hence our estimate would vary depending on the row chosen. There might even be one mean which was particularly different, for example if the rugby team were all sitting together in the back row.

Sampling error

An estimate from a sample (such as a mean) will not be exactly the same as the population value being estimated; in other words the estimate will be 'in error'.

'Error' here does not imply carelessness, but is an artefact of the method used to select the sample. There may be some bias involved in the sampling method. For example, choosing a sample from a telephone directory will systematically exclude those without phones and people who are ex-directory. In fact, prior to the 1936 US presidential election, a magazine called *Literary Digest* contacted 10 million voters using the telephone directory and lists of car owners asking about their voting plans. Based on 2 million responses the magazine forecast a comfortable victory for Landon over Roosevelt. However, Roosevelt ended up winning 61% of the votes. The problem was that in the 1930s in America or anywhere else poor people did not tend to have phones or cars and were thus under-represented in the sample. These people were apparently more likely to vote for Roosevelt. Sampling error can be reduced by using an optimum method of choosing a sample.

Non-sampling error

Non-sampling errors are other errors, not connected with the sampling method. For example:

- Poor survey questions may be interpreted incorrectly by respondents. For instance, in the 1949 Royal Commission of Population the following question was asked: 'Has it happened to you that over a long period of time, when you neither practised abstinence nor used birth control, you did not conceive? Yes/No.' In this example, it is unclear what timeframe is referred to and whether the frequency of intercourse deemed to be of any relevance.
- Interviewers can unconsciously bias the results, for example by their tone of voice or mannerisms. For instance, if an interviewer raises their voice and folds their arms in response to a particular view expressed by a respondent it may cause the respondent to moderate the strength of their responses in relation to this particular topic in future questions.
- Errors can be made when typing in or coding responses.

There is a huge amount of statistical theory devoted to estimating sampling error and recommending optimal sampling strategies which minimise error. If in doubt, contact an expert!

There are two main types of sampling: probability and non-probability. Probability samples are the best way of achieving random samples which are representative of the population, although even with this type of sampling it is unlikely that the sample will be exactly representative. They are also the main focus of the book. However, it is also important to introduce you to different forms of non-probability sampling, which don't involve *random* selection. You never know when you might need them!

Probability samples

There are four main types of probability sampling strategies.

Simple random sample

In a **simple random sample**, every member of a population has an equal chance of being selected. This is the sampling equivalent of the National Lottery, where we put all the members of the population into a hat and draw out their names. One of the problems with simple random sampling is that it requires a suitable sampling frame. While these are likely to be available for some populations (organisations such as unions, schools, or universities), adequate lists are not commonly available for larger populations such as a city or country.

Systematic sample

Systematic sampling is similar in many respects to simple random sampling. It is a form of random sampling but it involves a system. From the sampling frame a starting point is identified at random and thereafter at regular intervals. Systematically choose the xth individual from the frame, and continue this process until you have your entire sample. So if you wanted to sample 10 households on a street with 150 houses you would choose every 15th house after a random starting point between 1 and 15. I have known students employ this kind of approach when they have gathered too much data to deal with in a given timeframe, particularly with open-ended questions.

Stratified sample

A **stratified sample** is used if we know there are specific groups in a population who may be different from each other. For example, if we are interested in the proportion of MPs voting for entry into the euro, we know that certain groups of MPs may vote in particular ways, so we want to make sure we include them all. In other words, we want the sample to reflect the various strata in the population.

The most commonly used stratification in many samples is by sex and age. This is because as you get older you often change your views (Julie would say you get more conventional, but Ian denies this!) and men often have different views than women.

How many members of each stratum should be in a sample?

- The number is usually (but not always) proportionate to the number in the population. For example, if you wanted to look at attitudes towards smoking and you knew there were 200 smokers and 800 non-smokers working in a factory, you would want your sample of $n = 100$ to represent these groups fairly and would randomly select 20 smokers and 80 non-smokers.
- It is often a good idea to include larger numbers from the more variable and therefore less predictable strata.

Cluster sample

When it is necessary to have a sample from a population that's distributed across a wide geographic region, such as the whole of the UK, and it needs to be collected in person, you will have to cover a lot of ground geographically in order to get to each of the units you sampled. It is for precisely this dilemma that **cluster sampling** was developed. Cluster sampling involves dividing the population into clusters, randomly sampling clusters and measuring all units within sampled clusters. Using simple random sample techniques would result in covering the whole of the UK. In cluster sampling the researcher only randomly selects a number of clusters from the collection of clusters of the entire population. So if you were conducting research about universities in the UK you could randomly select a small number of universities (say, 15) from the entire population of universities and then interview 150 random selected students at each of the 15 universities. So, only a number of clusters are sampled; all the other clusters are left unrepresented. This is beneficial in terms of time, but certain subgroups of the population are excluded.

Non-probability samples

There are four main types of non-probability sampling strategies.

Convenience sample

Convenience sampling is the most common type of non-probability sampling. A convenience sample is one that is simply available to the researcher as a result of its accessibility. The issue with such approaches is that there is no evidence that they are representative of the population we want to generalise to. For instance, if

I wanted to research young adults I could simply ask students in my classes. This would be problematic in terms of the sampling frame as it would be biased in terms of education, social class, ethnicity and gender. However, it is quick to use this kind of approach.

Snowball sample

In certain respects **snowball sampling** is a form of convenience sampling. You begin by identifying someone who meets the criteria for inclusion in your study. You then make contact and ask them to recommend others they may know who also meet the criteria. Although this is unlikely to lead to a representative sample, there are times when it may be the best method available where access is problematic. Snowball sampling is especially useful when you are trying to reach populations that are inaccessible or hard to find (such as prostitutes, drug dealers or burglars). For instance, Becker (1963) used this approach in order to generate a sample of marijuana users in his classic study, initially making contact with people he had met as a professional musician, who he knew used marijuana and were able to put him in contact with other people who had used marijuana.

Quota sample

In **quota sampling** participants are selected non-randomly according to some fixed quota. It is used to provide a sample that reflects a population in terms of relative proportions of people in different categories such as gender, age or ethnicity. Deciding on the quota can be difficult and may rely on using data sources such as the Census to understand national characteristics. Unlike a stratified sample, where the sampling of the individuals is carried out randomly, the final selection of whom to include is down to the interviewer. So, if you know the population has 40% women and 60% men, and that you want a total sample size of 100, you will continue sampling until you get those percentages and then you will stop. This approach has led to criticism from those who use probability sampling approaches, as they argue it is not representative. However, it can be quicker than probability sampling alternatives.

Purposive sample

Purposive sampling involves sampling with a *purpose* in mind. It involves selecting a sample based on knowledge of a population, its subgroups, and the purpose of the study, selecting people who would be 'good' informants. It is commonly used in research carried out on the street by people with a clipboard. They might be looking for females between 30 and 40 years old to ask about a particular topic. The process involves looking at the people passing by and if anyone looks to be in that category they are stopped and asked if they want to participate. One of the first things to do is ensure that respondents meet the criteria for being in the sample. This is one of the weaknesses of this approach. For instance, this can be a little tricky in relation to age where people don't always want to say or your judgement is incorrect! However, it can be a very useful approach when you need to reach a targeted sample quickly.

How large should a sample be?

This is a tricky question! The sample size will reflect the size of a population, but will *not* usually be proportional to the population size. As a population gets larger, the sample size needed for an accurate estimate will not get proportionately bigger after a certain point.

The choice of sample size will depend on the variability among members of the population, the level of precision needed for the estimate, and the cost of sampling.

Variability among members of the population

Populations with greater variability (a high standard deviation) will need a larger sample size. If the population is very uniform, a small sample will be adequate for obtaining an accurate estimate. To take this to an absurd extreme, if everyone in the population were exactly the same then you would only need to sample *one* person to get a 'perfect' estimate of whatever you were trying to estimate!

Level of precision needed for estimate

The greater precision that is needed, the larger the sample that should be used. A larger sample will give a more precise estimate. It is worth noting here (as we shall see later) that the precision does not increase uniformly with sample size. After a while the increase in precision for a constant increase in sample size begins to diminish. It is worth noting that in clinical trials the sample is generally rather large in order to improve the representativeness in relation to the population. Furthermore, in order to undertake some statistical tests appropriately a sufficiently large sample size is also required.

Cost of sampling

Sampling can be very time-consuming and expensive. At the end of the day you only collect data if you really need to. Therefore the sample size should not be unnecessarily large.

The sampling distribution of the mean

Don't be put off by the title of this section: it's not as bad as it sounds!

We have already said that if we take lots of samples from a population, the means of each sample will be slightly different.

Suppose we take five samples A to E, each of five British women, and ask them how many children they have. The hypothetical data are shown in Table 8.1.

We could then calculate the mean number of children for each sample, as in Table 8.2. For example, the mean number of children in sample A will be:

$$\text{Mean} = \frac{0+2+2+1+1}{5} = \frac{6}{5} = 1.2$$

129

Table 8.1 Hypothetical data for five samples of five women who were asked how many children they had

Sample	Number of children				
A	0	2	2	1	1
B	3	0	1	0	2
C	2	1	2	2	0
D	2	0	4	1	0
E	1	3	0	2	2

Table 8.2 Mean number of children for the five samples of women

Sample	Mean number of children
A	1.2
B	1.2
C	1.4
D	1.4
E	1.6

The means of the different samples are called the **sample means.**

If the samples are representative, most sample means will be very close to the true population mean. However, a small number of samples will have a mean which is some distance away from the population mean. This leads us to:

Amazing fact number 1

The distribution of sample means will approximately follow the normal distribution.

Figure 8.1 shows the distribution of sample means.

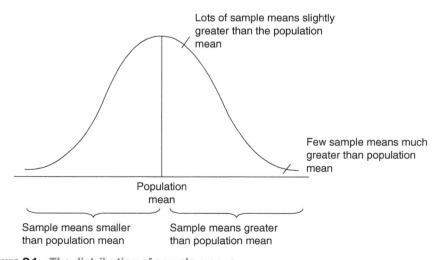

Figure 8.1 The distribution of sample means

If we have taken a reasonable number of samples, then:

Amazing fact number 2

The mean of the sample means will approximately equal the true population mean.

If you are getting confused with all these different means, look back at the example of the women and their number of children. The five means in Table 8.2 are the sample means. To find the *mean* of these sample means, we add the five means together and divide by 5:

$$\text{Mean of the sample means} = \frac{1.2 + 1.2 + 1.4 + 1.4 + 1.6}{5} = 1.36$$

So the mean of the sample means is 1.36 children, and from amazing fact number 2 we can therefore say that the population mean will probably be around 1.36 as well. In other words, the best estimate of the mean number of children for all British women is 1.36.

Now consider the sample means: will they vary as much as the actual values in the population? In Figure 8.2, which line shows the distribution of all the values in the population and which shows the distribution of the sample means?

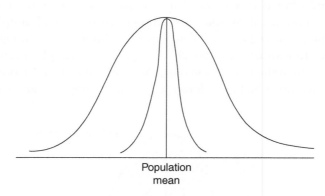

Population
mean

Figure 8.2 Distributions of population values and of sample means

The distribution with the much wider spread is the distribution of the whole population. The distribution with the much narrower range is the distribution of sample means.

Why is this?

The values for the whole population will include a few extreme values which make the range wider. But the sample *means* are means! Therefore in calculating them you will have effectively got rid of the extreme values.

To illustrate this, if the sample of five women had 2, 0, 1, 1 and 9 children respectively:

- The population distribution would include the extreme value of 9.
- The sample mean would still only be 2.6, which is not very extreme at all. Taking a mean moderates the effect of extreme values.

So the sample means are less variable than the actual values in the population. This will mean that the standard deviation of the sample means is smaller than the standard deviation of the whole population.

There is, in fact, a formula for calculating the standard deviation of the sample means. The standard deviation of the sample means is known as the **standard error** (SE):

Amazing fact number 3

$$SE = \frac{SD}{\sqrt{n}}$$

In words, the standard error is equal to the standard deviation of the population divided by the square root of the sample size.

We call the standard deviation of the sample means the standard error (SE) to distinguish it from the ordinary standard deviations of samples or populations.

The standard error is very handy. For example, since we know that the sample means follow a normal distribution, we can say that 68% of all sample means will lie between the mean and one standard error on either side, as in Figure 8.3. This idea enables us to estimate the chances of a particular sample mean being much larger or smaller than the population mean. We can calculate a Z-score for a sample mean to see how much higher or lower it is than the population mean.

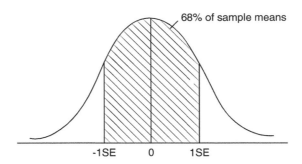

Figure 8.3 Distribution of sample means showing the standard error

Before looking at some examples, let us now combine the three amazing facts: together they are known as the **central limit theorem** (for no apparent reason!).

Fantastic theorem number 1: the central limit theorem

If samples of size *n* are selected from a population, the means of the samples are approximately normally distributed with

Mean = population mean

and

$$\text{Standard error} = \frac{SD_{pop}}{\sqrt{n}}$$

How big a sample do we need for this to work?

- If the distribution of the population is normal, a sample size of about 10 or more is needed.
- If the distribution is not normal (if it is skewed, for example), the sample size must be at least 25.

What use is all this? It is mainly useful for seeing whether a sample mean that you have found is different from the population due to chance (sampling variability) or whether the sample is genuinely quite different from the population. It also allows you to estimate the precision of any sample you take. The examples to follow show some applications of the theorem. These methods are the basis for hypothesis testing, which is introduced more formally in Chapter 11.

EXAMPLE: THE IQ OF STUDENTS

IQs in the general population are normally distributed with a mean of 100 and a standard deviation of 15. You take a random sample of 40 students and find that their mean IQ is 107. Given that you are prepared to accept that IQ measures intelligence, are the students really brighter than the average population?

What we are really trying to find out here is the chance of a sample of 40 students having a mean IQ of 107 or more due to chance (rather than genuine higher intelligence). This can be shown on a diagram, as in Figure 8.4.

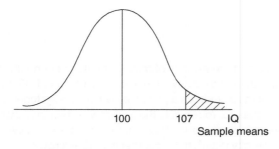

Figure 8.4 Probability of a sample mean of 107 or more

First we must find out the sampling distribution for the means of samples of size 40. According to the central limit theorem, the sample means will be normally distributed with:

$$\text{Mean} = \text{population mean} = 100$$

$$\text{Standard error} = \frac{\text{SD}}{\sqrt{n}} = \frac{1.5}{\sqrt{40}} = 2.37$$

Next we can calculate the Z-score using the formula:

$$Z = \frac{\text{sample mean} - \text{population mean}}{\text{standard error}}$$

$$= \frac{107 - 100}{2.37} = 2.95$$

A Z-score of 2.95 is quite high. Looking up $Z = 2.95$ on the normal table (Table A1.1) gives an area of 0.4984 between the mean and $Z = 2.95$. So the area above $Z = 2.95$ will be $0.5 - 0.4984 = 0.0016$.

This tells us that the probability of getting a sample of 40 people with an IQ of 107 or more just by chance is very low (less than 1%). Therefore we can conclude that we have evidence that the students genuinely are of higher intelligence than the general population.

Note that although the chance of getting a *sample* of 40 students with an IQ of 107 or more is extremely low, the chance of getting an *individual* with an IQ of 107 or more is much higher, because individuals vary much more than sample means. The Z-score for an individual with a score of 107 would be:

$$Z = \frac{\text{observation} - \text{mean}}{\text{SD}} = \frac{107 - 100}{15} = 0.47$$

Looking up a Z-score of 0.47 on the normal table gives an area of 0.1808. Therefore the area above $Z = 0.47$ will be $0.5 - 0.1808 = 0.3192$. So the chance of an individual having an IQ of 107 or more is 0.3192 (around 32%), which is quite reasonable.

EXAMPLE: THE WEIGHT OF STUDENTS

You have been told that the weights of male students are normally distributed with a mean of 70.31 kg and a standard deviation of 2.74 kg. You decide that your lecture class is representative of male students and that, by weight, male students will sit randomly around the lecture theatre. You sample the back row and when you weigh the nine men sitting there, their mean weight is 72.22 kg. Could this result have occurred due to chance, or are the rugby team sitting together in the back row?

According to the central limit theorem, the mean weights of samples of nine men will be normally distributed with:

$$\text{Mean} = \text{population mean} = 70.31 \text{ kg}$$

$$\text{Standard error} = \frac{2.74}{\sqrt{9}} = 0.91 \text{kg}$$

What is the probability of having a sample mean of 72.22 kg or more? This is the shaded area in Figure 8.5.

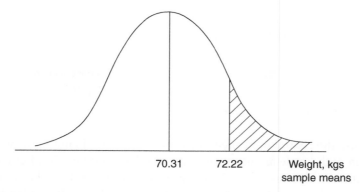

70.31	72.22	Weight, kgs
		sample means

Figure 8.5 Probability of a sample mean of 72.22 kg or more

$$Z = \frac{72.2 - 70.3}{0.91} = 2.09$$

The area between the mean and $Z = 2.09$ is 0.4817 (from the normal table, Table A1.1), and so the area above $Z = 2.09$ will be $0.5 - 0.4817 = 0.0183$.

The probability of a sample of men from the ordinary population of male students having a mean weight of 72.22 kg or more by chance is 0.0183 or 1.83%; in other words, the probability is very small. We must conclude that there probably is an abnormal group of people sitting in the back row, presumably the rugby team (are they scared, or what?).

EXAMPLE: BOTTLE FILLING

An EU rule states that wine producers must give you 750 ml in every bottle, otherwise they get grief from Brussels. Unfortunately, bottle filling machines are subject to variability. To be on the safe side, Monsieur de Vin sets his bottling machine up to put 755 ml of wine in each bottle. The machine has a natural variation which gives a standard deviation of 18 ml.

The next day, Madame Bruxelles, the inspector, comes along and does a random check of 36 of Monsieur de Vin's bottles. What is the chance of Monsieur de Vin getting grief from Brussels?

What we want to know is the proportion of samples which will contain less than 750 ml of wine on average, because Monsieur de Vin would be in trouble if the mean of his sample was less than 750 ml. The proportion is the shaded part in Figure 8.6. (Always draw a diagram to help with these questions.)

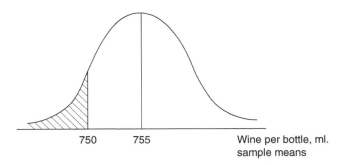

750 755   Wine per bottle, ml.
                 sample means

Figure 8.6 **Probability of grief from Brussels ...**

We can use the central limit theorem, even though we do not know whether the quantities in each bottle are normally distributed, because $n = 36$. Remember that the population in this case (by which we mean this example, rather than 12 bottles!) is all of Monsieur de Vin's bottles, and a sample is any 36 randomly selected bottles.

According to the central limit theorem, samples of 36 bottles will be normally distributed with:

$$\text{Mean} = \text{population mean} = 755 \text{ ml}$$

$$\text{Standard error} = \frac{\text{SD}}{\sqrt{n}} = \frac{18}{\sqrt{36}} = 3 \text{ ml}$$

Now that we know the standard error, we can calculate Z:

$$Z = \frac{750 - 755}{3} = -1.67$$

From the normal table, the area between the mean and $Z = -1.67$ is 0.4525. So the area below $Z = -1.67$ will be $0.5 - 0.4525 = 0.0475$. Therefore the chance of Monsieur de Vin's bottles failing the test is 0.0475 or 4.75%.

What would you do? Risk it? Buy a more reliable machine? Increase the amount the machine should give? Any of these strategies might be successful. In statistics, we often use a 5% risk as a cut-off point. In this example, the chance of the sample failing is less than 5%, so Monsieur de Vin can be over 95% certain that his sample will pass. If he wants to be more confident, he must reset his machine to put more wine in each bottle (but this will increase costs). As in life, it is a question of weighing up the risks (and punishments!), costs and benefits of different strategies.

Guidelines for using the sampling distribution for the mean

From the three examples, it should be clear that the strategy for answering questions involving the sampling distribution for the mean is:

1. Find the mean and standard error of the sample means.
2. Draw a diagram.
3. Calculate the Z-score.
4. Look up the probability on the normal table and draw a conclusion.

Summary

This chapter has introduced you to sampling and how this needs to be as representative of the population as a whole as possible. It has also introduced you to different forms of sampling and common problems with sampling strategies. It indicates that sampling can be a rather complex process which has important consequences for the kind of generalisations we can make. For instance, if we use a convenience sample of betting habits in a betting shop in Sheffield these are not necessarily representative of behaviour in all betting shops, because the area that shop is in may be characterised by particularly high or low wages. The chapter has also shown how the use of samples may not only be less time-consuming but also enable research to take place which wouldn't be possible if a whole population had to be used. In addition, the use of sample means has been looked at, including how these can be used to predict particular outcomes in social research.

PRACTICE QUESTIONS

8.1 Casual workers in a particular occupation are paid a mean wage of £6.60 per hour with a standard deviation of £0.40. The wages are normally distributed.

(a) What is the probability of an individual worker earning £6.50 or less per hour?
(b) What is the probability of obtaining a sample of 20 workers with a mean wage of £6.50 or less?
(c) What is the probability of obtaining a sample of 50 workers with a mean wage of £6.50 or less?
(d) Why are your answers to parts (a), (b) and (c) different?

8.2 Casual workers in another industry are paid on average £8.10 per hour with a standard deviation of £0.50. Workers in a particular firm believe that they are being underpaid. A sample of 30 casual workers from the firm has a mean hourly wage of £6.50. Is this low mean wage likely to be due to chance or are the workers in this firm really hard done by?

Hints: find the mean and standard error of the sample means. Draw a diagram and find the Z-score for the sample. What is the probability of getting a sample with a mean of £6.50 or less by chance?

8.3 The mean age of mothers at the birth of a child in England and Wales is 29.8 years (Office for National Statistics, 2013d). Suppose that the standard deviation is approximately 4.6 years. You believe that women in a deprived area are having their births at a younger age than average. You take a sample of 100 births from the local hospital and find the mothers' ages at the time of the birth. The mean age in the sample is 29.0 years. Are these women really having births at a younger age or could the difference be due to sampling variability?

NINE
GETTING CONFIDENT

Introduction

A statistical question often asked is how to estimate the mean of a population from a sample. For example:

- You want to take a sample of students in a Quantitative Methods lecture so as to estimate the mean height of all the students on the course
- A sociologist wants to know how much income women lose or gain on divorce, so they interview a sample of divorced women

In these cases, the strategy is to take a representative sample and then estimate the population mean from the sample. This process involves the use of **confidence intervals**. A confidence interval provides a range within which we can have a degree of confidence that the estimate is not due to chance. This chapter discusses how confidence intervals are calculated and their presentation in the form of **hi-lo plots**. By the end of the chapter you should be able to:

- Understand when it may be appropriate to estimate a population mean from a sample
- Calculate a confidence interval and interpret it
- Present confidence intervals in the form of a hi-lo plot

Is a sample mean a good estimate of the population mean?

Do you think that a sample mean would be a good estimate of the population mean?

Suppose the mean height of 30 people in the Quantitative Methods lecture is found to be 171 cm. This sample mean will not be exactly the same as the mean height of the whole class (the population mean), but it might be a good estimate.

An estimate of the mean like this is called a **point estimate**: we are trying to pin the mean down to an exact figure.

A sample mean is the best point estimate for the population mean.

But, you may be thinking, how can a mean of any sample be a good estimate of a population mean when we know that if we take repeated samples from the same population we will get a different answer for the mean every time? The answer is to remember the results from the last chapter, which showed that properly collected samples vary in a systematic way. Also bear in mind that sampling error can have significant implications for the usefulness of working out a population mean from a sample. The point estimate is the best 'one-number' estimate, but what we would really like, in addition, is an estimate of the likely range in which we can be reasonably sure the true population will lie. This is called an **interval estimate** and is intuitively better than a point estimate.

For example, most people would be happier, if asked to estimate the number of students in a lecture, to say something like 'between 50 and 60' or 'around 50' than to estimate a precise number such as 55. When estimating a population mean, the way to do this is to calculate a confidence interval.

Confidence intervals

You can make estimates for any level of confidence you like. However, most often people find the range within which we are 95% certain the mean will lie: this is called a 95% confidence interval. In other words, we will calculate a range in which we would expect the true value to lie 19 times out of 20. While the 95% confidence interval is commonly used, the percentage you choose depends on the research you are conducting and personal preference.

Suppose that the mean height of the 30 students is 171 cm with a standard deviation of 12 cm. We want to calculate an interval estimate for the mean of all the students in the room.

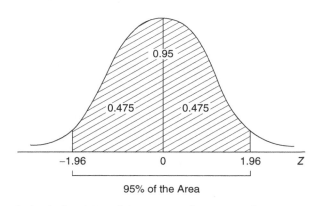

Figure 9.1   Graph depicting 95% of the area under a normal curve

Think about a normal distribution. If we look at the 95% in the middle of the distribution, it can be broken down into two halves, with 47.5% of the area on either side of the mean, as in Figure 9.1.

If you now look in the *middle* of the normal table (Table A1.1) for the area 0.475, you will find that it corresponds with a Z-score of 1.96. In fact, 1.96 is the magic number that you need to remember for calculating 95% confidence intervals. To work out the possible variation around the sample mean we multiply the standard error by 1.96:

$$1.96 \times \frac{SD}{\sqrt{n}} = 1.96 \times \frac{12}{\sqrt{30}} = 4.29 \text{ cm}$$

Thus we have a range spanning 4.29 cm on either side of the sample mean where we are 95% confident that the population mean will lie.

To work out the two limits between which the population mean could lie, we need finally to add and subtract this figure from our sample mean:

$$171 + 4.29 = 175.29$$

$$171 - 4.29 = 166.71$$

Therefore we can be 95% confident that the population mean will lie between 166.71 cm and 175.29 cm. This is usually written as follows, with the lower limit first:

$$95\% \text{ CI} = (167, 175)$$

The formula for a 95% confidence interval is therefore:

$$95\% \text{ CI} = \text{sample mean} - \left(1.96 \times \frac{SD}{\sqrt{n}} \right)$$

A confidence interval therefore takes into account both the variability of the sample and the size of the sample. A sample which is larger or less variable will produce a better estimate of the population mean and so the confidence interval will be narrower than it would be for a smaller or more variable sample.

Note that to use this formula we need to remember that fantastic central limit theorem. Either the population must be normally distributed or the sample size must be at least 30. Strictly speaking, the standard error should use the population standard deviation rather than the sample standard deviation, but for confidence intervals we can use the sample standard deviation, provided that the population is not too small.

EXAMPLE: FREQUENCY OF INTERCOURSE

Let's return to the data on frequency of sex among married women aged 19–20 in the USA (from Table 5.2). Table 9.1 gives the mean frequency of intercourse in the past four weeks, along with the standard deviation and sample size for each of the three years. The question was whether there was an increase in the frequency of sex over time.

Table 9.1 Frequency of intercourse in four weeks before interview for married women aged 19–20 in the USA, 1965, 1970 and 1975

	Year of interview		
	1965	**1970**	**1975**
Mean	9.6	9.8	12.1
Standard deviation	1.1	1.3	1.2
n	190	249	219

Source: Means from Trussell and Westoff, 1980; other data hypothetical

Now that we know how to calculate confidence intervals for the three means, we can answer this question properly. For example, the 95% confidence interval for 1965 would be:

$$95\% \, CI = 9.6 \pm \left(1.96 \times \frac{1.1}{\sqrt{190}} \right) = 9.6 \pm 0.16$$
$$= (9.44, 9.76)$$

In other words, we can be 95% confident that the true value for frequency of intercourse in the population of married women aged 19–20 lies between 9.44 and 9.74 times in a four-week period.

See if you can calculate the 95% confidence intervals for 1970 and 1975 yourself and check that you come up with the following answers:

1970: 95% CI = (9.64, 9.96)

1975: 95% CI = (11.94, 12.26)

What do you notice about the confidence intervals around the three means?

The confidence intervals for 1965 and 1970 overlap. However, the confidence intervals for 1970 and 1975 do not overlap. We should only conclude that there is a real difference if the confidence intervals do not overlap. Therefore the data show no real difference in the frequency of sex among American married women aged 19–20 between the years 1965 and 1970. There is evidence of an increase in frequency, however, between 1970 and 1975.

As we have already noted, we can use confidence intervals other than 95% if we want to, although 95% is the most common. For more precision, we might want to use a 99% CI. Can you work out what the formula for a 99% CI would be? (Hint: only the 1.96 would change.)

It would be:

$$99\% \, CI = \text{sample mean} \pm \left(2.575 \times \frac{SD}{\sqrt{n}} \right)$$

Half of 99% is 49.5%, as in Figure 9.2. The figure of 2.575 was found by looking up 0.495 on the normal table and finding it to be halfway between 2.57 and 2.58.

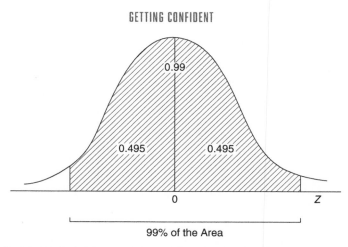

Figure 9.2   Graph depicting 99% of the area under a normal curve

Note that the confidence interval gets wider as the precision increases. Why is this? Put simply, if you want to be ever more certain that the true value will lie within your interval estimate then you will need a wider interval. On the other hand, if you are happy to be wrong most of the time (which would be rather odd!) then you can have a smaller interval. There is nothing sacrosanct about 95%: it is just a range which has widespread intuitive acceptability among analysts.

Hi-lo plots

Confidence intervals can be presented graphically by using a hi-lo plot. A hi-lo plot shows the estimated mean and the confidence interval for one or more populations. They are particularly useful for comparing the means and confidence intervals for several populations side by side. For each population, the length of the vertical line represents the range of the 95% (or whatever value you choose) confidence interval and the middle horizontal line is the mean.

EXAMPLE: DRINKING BEHAVIOUR

In order to understand drinking behaviour with the view to informing health policy a researcher wanted to know if there were daily variations in alcohol consumed on different days and whether this represented a binge culture. In order to do this they found information about the number of pints of beer sold in 10 pubs in their area over a 12-week period on different days of the week.

The results are shown in Figure 9.3. There does appear to be some daily variation in the number of beers purchased, with a higher mean on Fridays and Saturdays. The 95% confidence intervals for the days Sunday to Thursday all overlap, so we cannot make any concrete conclusions about daily differences in the

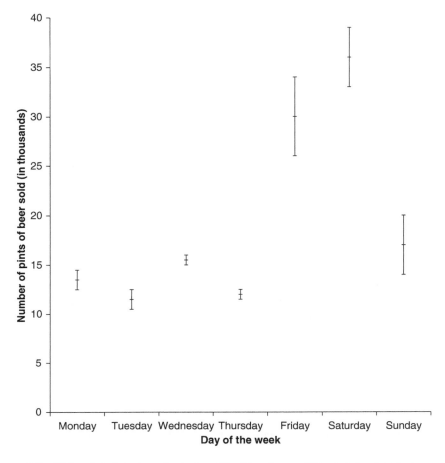

Figure 9.3 Hi-lo plot showing alcohol sold in 10 pubs on different days over a 12-week period

Source: Hypothetical data

number of pints of beer sold in 10 pubs over a 12-week period for these days. However, the confidence intervals for Fridays and Saturdays do not overlap with the other days of the week, so we can say that there is a significant difference in the number of pints of beer sold in the 10 pubs over a 12-week period between Fridays and Saturdays and the rest of the week.

Do you think that counting the number of beers sold in pubs is a good method of measuring alcohol consumption on various days?

There are a number of issues:

- Does it take into account the number of people buying the beers?
- What about other drinks sold, or half pints?
- Are the 10 pubs representative of pubs nationally?
- What about alcohol purchased and consumed in a different setting, such as at home?
- Was all the beer sold actually consumed? (Just because it was sold doesn't mean it was necessarily drunk; we all hate wasted beer though!)

It is impossible to gauge how much individuals are drinking from the data, just the number of beers sold in 10 pubs on particular days. However, this information is likely to be useful for the landlords' knowledge on when to order more beer if they don't normally keep records!

Summary

This chapter has introduced you to the process of estimating the mean of a population based on a (suitable) sample. For instance, if you had information about the time it took for funeral payment claims to be paid in a small sample of local authorities, how could this be used to estimate the mean time local authorities take to process funeral payment claims nationally? Questions like this can be addressed through the use of confidence intervals. You should now be aware of how these are calculated and how to present them meaningfully in the form of hi-lo plots.

PRACTICE QUESTIONS

9.1 Vauclair et al. (2010) conducted some research about attitudes towards older age. Four hundred people participated in the study: 200 students (the younger age group) who had an average age of 17.55; and 200 people over 55 (the older age group) with an average age of 70.16. They were asked, 'At what age do you think people generally start being described as old?' and the findings are shown in Table 9.2.

(a) Does it appear that the mean age at which respondents think people generally start being described as old varies by their own age at the time of being questioned?

(b) Calculate 95% confidence intervals for the mean age at which respondents feel people generally start being described as old for the younger and older age groups. What can you conclude now about any real variations in attitudes towards older age?

Table 9.2   Age in years at which respondents think people generally start being described as old

	Age group	
	Younger group	**Older group**
Mean	53.15	66.00
Standard deviation	11.25	9.26
n	200	200

Source: Vauclair et al., 2010

9.2 A sample of 40 judges were asked their opinion on the death penalty, on a scale of 1 to 10, where 1 is 'totally disagree' and 10 is 'totally agree'. The judges' mean score was 3 with a standard deviation of 5.5.

(Continued)

(Continued)

(a) Estimate the 'death penalty rating' for all judges in the UK and calculate a 95% confidence interval for your answer.

(b) Calculate a 99% confidence interval for your answer. Is this wider or narrower than the 95% confidence interval? Why?

(c) Suppose that you had obtained the same mean and standard deviation from a sample of 100 judges. Calculate a new 95% confidence interval for the 'death penalty rating' for all the judges in the UK. Is this wider or narrower than your answer to part (a)? Why?

9.3 You are investigating housing availability for lone parents with one child in your local area. To get an idea of the price of houses for sale, you look in an estate agent's window and note down the prices of their two-bedroom houses on offer. Table 9.3 gives these prices.

Table 9.3   Prices of two-bedroom houses at an estate agent (£)

30,000	40,500	79,950	101,000	145,900	72,250
172,500	135,950	183,600	91,450	255,500	

Source: Hypothetical data

Estimate the mean price of a two-bedroom house in the area and give a 95% confidence interval for your answer. Do you think that this sample of prices is representative?

TEN

FUN WITH PROPORTIONS

Introduction

In the last three chapters we have been looking at continuous variables, where different samples and populations can be compared by examining their means. Sometimes the statistic of interest will be in the form of a proportion rather than a mean. For example, you might be interested in the proportion of senior staff in a company who are female or the proportion of teenagers who regularly take hard drugs. Confidence intervals can also be used when working with proportions, although there are differences in how these are calculated. This chapter will introduce you to the process of using proportions to work out confidence intervals. By the end of the chapter you should be able to:

- Understand how to use proportions to produce confidence intervals
- Evaluate when it is useful to use them in this manner
- Identify how to evaluate the findings

Proportions

If you wanted to know what proportion of students believe that Britain was right to go to war with Iraq in 2003 you would take a representative sample of students and ask them if they thought Britain was right to go to war with Iraq. You might group people's answers into 'yes' or 'no or don't know'. Let's assign a value 1 to 'yes' and 0 to 'no or don't know'.

The proportion of those who thought Britain was right to go to war with Iraq will be:

$$\text{Proportion} = \frac{1+1+1+\ldots+0+0+0}{\text{total number in sample}} = \frac{\text{number who thought Britain was right to go to war with Iraq}}{\text{total number in sample}}$$

The question is: can we use this as an estimate of the proportion of the whole population (all students) who thought Britain was right to go to war with Iraq?

The good news is – yes we can!

Some jargon

- We call the sample proportion p.
- The true population proportion we are trying to estimate is known as Π. (Π is the Greek capital letter 'pi', pronounced 'pie'.)

Now it's time for another magnificently useful and fantastic theorem! This is like the central limit theorem but refers to a *proportion* rather than a mean.

Fantastic theorem number 2

Provided that n is reasonably large, the distribution of sample proportions p is approximately normal with

mean $= \Pi$

and

$$\text{Standard error} = \sqrt{\frac{\Pi(1-\Pi)}{n}}$$

This is similar to the distribution of sample means. What it implies is that if we take repeated samples and find the proportion p of each sample, then the mean of all the ps will be approximately the population proportion Π, and the standard deviation of the distribution of ps will be the standard error given in the equation.

As long as the sample size is reasonably large, we can use the sample proportion p as a reasonable approximation to the population proportion in the standard error. Here are some examples putting 'Fantastic theorem number 2' into practice.

EXAMPLE: MOT TESTS

A few years ago, in 2008, your car, a nice BMW 3-Series (2003 model), had just failed its MOT, but you thought that it was perfectly OK and the garage was making things up. On checking the garage's records you found that, out of 100 MOT tests on BMW 3-Series (2003 models) in that year, 55 resulted in a fail. Therefore the sample proportion is 0.55 from a sample where $n = 100$.

National figures state that, in 2008, 36% of MOT tests of 55,621 BMW 3-Series (2003) cars resulted in a fail (Department of Transport, 2013). Thus the population proportion is 0.36.

You wanted to find out whether the garage's result of 0.55 is likely given that the population proportion is 0.36, or whether they are failing lots of cars unnecessarily to make more money.

According to 'Fantastic theorem number 2', the proportions failing in different samples of 100 will be normally distributed with:

$$\text{Mean} = \text{population proportion} = 0.36$$

$$\text{Standard error} = \sqrt{\frac{0.36(1-0.36)}{100}} = \sqrt{\frac{0.23}{100}} = 0.048$$

Now we need to find the chance of having 55 fail results in a sample of 100, so, in jargon, we want to find the probability that $p \geq 0.55$.

As we know p, Π and the standard error, we can put these figures into the equation to find a Z-score:

$$Z = \frac{\text{sample proportion} - \text{population proportion}}{\text{standard error}} = \frac{p - \Pi}{SE}$$

$$= \frac{0.55 - 0.36}{0.048} = \frac{0.19}{0.048} = 3.96$$

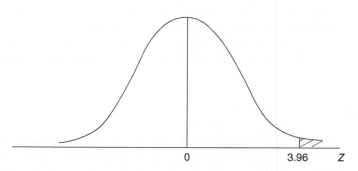

Figure 10.1   A Z-score of 3.96

Figure 10.1 shows this on a diagram. A Z-score of 3.96 is very high and is not on the normal table (Table A1.1), so the area greater than $Z = 3.96$ is extremely small. It is very unlikely that the garage would have failed 55 out of 100 MOTs just by chance. Either the customers had particularly dodgy cars (or ones with larger mileages) or the garage was fiddling things to make more money on fake repairs. I suggested you should try a different garage next time!

But what if the garage had only failed 40 out of 100 cars in the sample? Would this still have been a cause for concern?

The sample proportion p is now 0.40, and the population proportion and sample size remain unchanged. The mean and standard error of the sample proportions will therefore be the same. All we need to do is calculate a new Z-score:

$$Z = \frac{0.40 - 0.36}{0.048} = \frac{0.04}{0.048} = 0.83$$

149

The shaded area in Figure 10.2 shows the probability of getting a sample with a proportion of 0.40 or more. The area between the mean and $Z = 0.83$ is 0.2967 (from the normal table). Therefore the area above $Z = 0.83$ will be $0.5 - 0.2967 = 0.2033$.

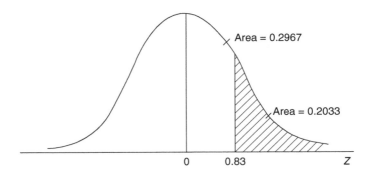

Area = 0.2967

Area = 0.2033

0   0.83   Z

Figure 10.2   A Z-score of 0.83

Therefore there was a 20% chance of the garage having a proportion of 0.40 or more of its MOT tests resulting in a fail. This is a reasonable chance and could be due to natural sampling variability. There would probably have been no need to be suspicious of the garage if only 40 MOT tests in the sample had been failed.

EXAMPLE: SOCIAL WORK BURSARIES

Following consultation in May 2013, the Department of Health-led Bursary Prioritisation and Advisory Group reported that the government had agreed to continue to provide bursaries to postgraduate Social Work students on MA courses (Department of Health, 2013). However, a cap on the number of bursaries for students entering such courses from 2014 was put in place, with different universities allocated different numbers of bursaries. Universities then need to decide how the bursaries are allocated. One option is to use a first come, first served approach in relation to bursaries for those candidates made an offer of a place on the course. So if a university had 50 bursaries it would offer these to the first 50 people to accept a place on the course. As the admissions tutor you know from previous experience that on average only 95% of people accepting a place on the course turn up at the start of the course. As a result of this you decide to operate a reserve list of a further five people who accepted their places later and want a bursary. One of the people on the reserve list contacts you prior to the start of the course and asks what their chance is of receiving a bursary at the start of the course. You have already allocated your 50 bursaries and they are the fifth reserve. How would you work this out?

Let's first write down what we know. We know that the sample size n is 50 and that the population proportion Π is 0.95 (95 ÷ 100).

We can then calculate the mean and standard error of the sample proportions:

$$\text{Mean} = \Pi = 0.95$$

$$\text{Standard error} = \sqrt{\frac{\Pi(1-\Pi)}{n}} = \sqrt{\frac{0.95 \times 0.05}{50}} = 0.097$$

So we can say that p is normally distributed with mean = 0.95 and SE = 0.097.

Next we must work out what must happen in order for them to get a place on the course with a bursary. There are 50 places with bursaries and they are the fifth reserve. Therefore for them to get a place with a bursary, how many people would have to not turn up? Clearly, if 50 – 5 = 45 of those entitled to a place with a bursary (or fewer) turned up you would receive a bursary. But what is the chance of so few people turning up?

The sample proportion p will be 45 ÷ 50 = 0.90. This is the maximum proportion of students you want to show up to ensure success. So you want to find the probability of the proportion turning up being 0.90 or less.

Now we can calculate the Z-score and draw a diagram (see Figure 10.3):

$$Z = \frac{0.90 - 0.95}{0.097} = -0.52$$

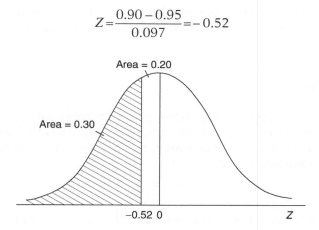

Figure 10.3 Probability of the fifth reserve student receiving a bursary

From the normal table (Table A1.1), the area between the mean and $Z = -0.52$ is 0.20, and so the area to the left of $Z = -0.52$ will be 0.50 – 0.20 = 0.30.

From this result you can conclude that the fifth reserve student has a 30% chance of getting a place on the course with a bursary.

Confidence intervals for proportions

You will remember that when we estimated a population mean from a sample we argued that, intuitively, an interval estimate was often preferable to a point estimate, and therefore we calculated a confidence interval as a measure of the range we could be sure that the population mean would lie in.

In the same way we can calculate confidence intervals for proportions, so that when we estimate a population proportion from a sample we can have some idea of how accurate our estimate is.

The formula for a 95% confidence interval is analogous (comparable in many respects) to that for a mean:

$$95\% \; CI = p \pm (1.96 \times SE)$$

So the actual equation will be as below (putting in the formula for the standard error of a proportion) using the sample proportion p – which in practice will be all we have, because if we knew what the population proportion was we wouldn't need to estimate it:

$$95\% \, CI = p \pm \left(1.96 \sqrt{\frac{p(1-p)}{n}} \right)$$

As with means, for a 99% confidence interval, multiply the standard error by 2.575 instead of 1.96. This will make the confidence interval wider, because the wider the range, the more confident you can be that the population proportion will lie in it.

EXAMPLE: PENSION PLANNING

Liam did some work with AXA Wealth about pension planning, conducting a survey of 200 current pensioners. Respondents were asked the question 'Did you retire when you expected to?' Out of the 200 pensioners who answered the question, only 58 replied 'yes' (AXA Wealth, 2012).

This gives us a sample proportion of 0.29 (58 ÷ 200). We can use this sample proportion p as an estimate of the population proportion Π, but we must have some measure of the variation in the sample proportions. In this case, a 95% CI would be:

$$95\% \, CI = 0.29 \pm \left(1.96 \sqrt{\frac{0.29(1-0.29)}{200}} \right) = 0.29 \pm 0.0629 = (0.2271, 0.3529)$$

Therefore we can say with 95% confidence that the population proportion will lie between 0.2271 and 0.3529. In other words, between 23% and 35% of pensioners retire when they expect to.

Confidence intervals for proportions are particularly useful when interpreting **opinion polls**.

Table 10.1 compares the results of two British opinion polls published in February 2012 which asked a sample of people how they would vote if there was a general election tomorrow. The results from the poll conducted by YouGov for the *Sunday Times* suggested that Labour was in the lead by 1%, while the results from the YouGov survey for the *Sun* suggested that the Conservatives had a 2% lead over Labour. Does this mean that Labour were in the lead or the Conservatives? Which results should we believe?

Table 10.1 Percentage of people intending to vote Conservative or Labour in two opinion polls held in February 2012

		% intending to vote:	
Opinion poll	Number in sample	Conservative	Labour
YouGov/*Sunday Times*	1,659	39.0	40.0
YouGov/*Sun*	1,763	40.0	38.0

Source: Original data from YouGov, 2013

As a good statistician, you are a little wary of the papers' claims, so you decide to calculate 95% confidence intervals for all the results.

The percentages must first be converted into proportions, so for example the 39% intending to vote Conservative according to the YouGov *Sunday Times* poll is a proportion of 0.39. The 95% confidence interval for the proportion voting Conservative would be:

$$95\% \, CI = 0.39 \pm \left(1.96 \sqrt{\frac{0.39(1-0.39)}{1659}} \right) = 0.39 \pm 0.0235 = (0.3665, 0.4135)$$

Converted back into percentages, this could be written as:

$$95\% \, CI = (36.65\%, 41.35\%)$$

Try the other three yourself and see if you come up with the results in Table 10.2.

If you look at the ranges of the confidence intervals, it is obvious that we cannot draw any concrete conclusions about who would win the elections from these data if they were held tomorrow. Because the ranges for the percentages voting Conservative and Labour overlap each other in each poll, we cannot say from the YouGov *Sunday Times* or YouGov *Sun* polls that either party was in the lead.

Table 10.2   Ninety-five per cent confidence intervals for the percentage of people intending to vote Conservative or Labour in two opinion polls held in February 2012

Opinion poll	% intending to vote (95% confidence interval)	
	Conservative	Labour
YouGov/*Sunday Times*	39.0 (36.65, 41.35)	40.0 (37.64, 42.36)
YouGov/*Sun*	40.0 (37.70, 42.30)	38.0 (35.73, 40.27)

Source: Original data from YouGov, 2013

The moral of the story is: don't believe everything you read about opinion polls, even in the 'quality' newspapers! In fact, in a similar example on the day prior to the 1992 general election NOP/*Independent* had Labour as the winner of the election by 3% and Gallup/*Telegraph* the Conservatives by 0.5%. In the end the Conservatives won with a 7.6% lead, which the polls had failed to predict (Butler and Kavanagh, 1992)! This was blamed on various factors such as late switching and differential turnout among Conservative and Labour supporters.

Summary

This chapter has built on the previous ones, considering how confidence intervals can be employed to make predictions about the population. However, this chapter has focused on the use of proportions in order to do this rather than means. You should now know how to use proportions in collaboration with confidence intervals to explore a population. This is really useful when thinking about the confidence we have in our conclusions.

PRACTICE QUESTIONS

10.1. You decide to take a representative sample of students – 50 males and 50 females – and ask them what they thought about one-night stands. You found that 21 of the males and 35 of the females responded by saying that one-night stands are usually not a good idea.

(a) You want to apply your results to the whole student population, so calculate the 95% confidence intervals for the proportions of male and female students who believe that one-night stands are usually not a good idea. Can you say that there is a definite difference in the views of males and females from your sample?

(b) Because the ranges of possible values for Π are quite large, you decide to rerun the survey with larger samples of 150 males and 150 females. This time you find that 65 males and 110 females believe that one-night stands are usually not a good idea. Now what figures can you be 95% confident that the population proportions will lie between?

10.2. Of the 1,548,000 new visitors to sexual health clinics in the UK in 2011 among people not previously diagnosed with HIV, 20% of people declined the offer of an HIV test (Health Protection Agency, 2012). If you take a random sample of 100 new visitors to sexual health clinics across the UK, what is the chance that:

(a) 30 or more people will decline the offer of an HIV test?
(b) 15 or fewer people will decline the offer of an HIV test?
(C) Between 23 and 28 people will decline the offer of an HIV test?

10.3. A local market research company takes a random sample of 200 adults in a town and asks them whether they support the council's plans for building a multiplex cinema complex on a greenfield site outside the town. Only 80 of the adults say that they support the proposal.

(a) Estimate the proportion of the adult population who support the proposal and calculate a 95% confidence interval for your answer.
(b) Calculate a 99% confidence interval. Explain why it is different from the 95% confidence interval.
(c) A member of the council claims that the market research company has got it wrong and that 60% of the population support the cinema proposal. Do you think the council member could be right?
      Hint: what is the chance of getting a sample of 200 people with the observed proportion supporting the proposal if the true population proportion is 0.6?

ELEVEN

HOW TO DECIDE HOW TO DECIDE

Introduction

We have already been using the normal distribution to find out the probability of something happening (see Chapter 8). In a similar way, it can be used for formal decision-making. This chapter and the next provide a brief introduction to **hypothesis** testing. This is an important concept in quantitative research and is commonly used when discussing research design. A hypothesis is used once you have developed your research interest, research aim, and some potential research questions. Hypotheses are designed to express relationships between variables and, crucially, whether there is a significant relationship between those variables. It is a statement that we are trying to prove or disprove.

Hypotheses can also be used to explore how a particular sample compares with the whole population. This is the main focus of this chapter. For instance, it is possible to examine a hypothesis in the light of sample evidence in the form of a significance test using confidence intervals. A more common approach, using the test statistic, will also be discussed where the sample mean is compared to the population mean assumed under the hypothesis using Z-scores. It will also show the difference between **two-sided** (or two-tailed) **tests** and **one-sided** (or one-tailed) **tests** in calculating whether a hypothesis should be proved or disproved. By the end of the chapter you should:

- Understand what a null hypothesis and alternative hypothesis are
- Know when it is appropriate to undertake hypothesis testing, and how to do this using confidence intervals, and Z-scores when you have sample means or proportions
- Be aware of the difference between two-sided (or two-tailed) tests and one-sided (or one-tailed) tests and when they should be used

Developing a hypothesis

Research hypotheses are essential if you are exploring two or more variables in conjunction with each other or you are looking at how a sample compares with a population. If your research questions are more descriptive then generating a hypothesis may not be appropriate. However, most quantitative research is **deductive** in nature – that is, particular theories are being tested rather than generated – so forming a hypothesis is particularly important.

We often want to know whether a sample is likely to come from a particular population or whether a sample is genuinely different from that population. For example, a large company had an 'away day' during which its employees had the opportunity to take free IQ tests. They all decided to participate. The employees (anonymously) then provided their employer with their IQ scores. These indicated that their staff were very clever, with professional staff having a mean IQ of 120, with a standard deviation of 10. The company decided that they thought some staff might be misleading them and presented this theory to the staff. The staff were amazed and all said they would be prepared to do it again to prove their intelligence. The company selected 36 staff at random to retake the test and provide the IQ test results on the official test score paper. They found that the mean IQ of the 36 staff was 114. It appeared that the staff tested had, on average, a lower IQ than the employees claimed. There are two possible reasons for this:

- The employees' claim is *true*: the staff do have a mean IQ of 120. By chance a sample of the staff was selected with lower IQs
- The employees' claim is *false*: the staff do not have a mean IQ of 120

Which explanation is likely to be the correct one? How do we decide what to decide?

First of all, we must write down a hypothesis about what we expect to happen if the employees' claim about their score is true. This is known as the **null hypothesis** or H_0. The null hypothesis is something of a 'life is boring and nothing exciting (or out of the usual) is happening' hypothesis. In this example, our null hypothesis is that the company staff do have a mean IQ of 120, so we could write:

$$H_0: \text{The company staff have a mean IQ of 120}$$

The question we are then asking is 'Do our sample data support this hypothesis?'

From Chapter 8, we know that any sample will not behave exactly as we would expect under our null hypothesis. Nevertheless, provided that our hypothesis is sensible, we would expect our sample value to fall in a region of values fairly close to the population value that is specified by the null hypothesis. For example, if you had found that the mean IQ in your sample was 119, you would probably be fairly happy to say that the population mean could be 120. However, faced with a mean IQ in the sample of 114, you would rightly be more sceptical of the truth of the null hypothesis. Where do we draw the line between believing and not believing the employees' claim regarding their initial scores?

If, as in this case, the sample value falls a relatively long way from our expectation, then we must ask:

- Is our hypothesis sensible?
- Is the sample result too much of a coincidence just to be a fluke? In other words, is it highly probable that some alternative hypothesis is true?

Secondly, we write down an **alternative hypothesis** (sometimes this is called the research hypothesis or the experimental hypothesis), denoted by H_A or H_1. In this case we would be thinking it more likely that the employees' claim is false, so we could write:

$$H_A = \text{the company staff do not have a mean IQ of 120}$$

Note that the population mean is usually written as μ. In this example the two hypotheses could be written more concisely as:

$$H_0: \mu = 120 \text{ (the sample comes from a population that has a mean of 120)}$$

$$H_A: \mu \neq 120 \text{ (the sample does not come from a population that has a mean of 120)}$$

Now we are faced with the really tricky problem. How can we decide whether to believe the null hypothesis or whether to go for the alternative hypothesis? Hypothesis testing is all about finding strategies to make this decision.

What we need is a range of possible sample values within which we will be happy to accept the null hypothesis, but if the population value lies outside that range we would decide not to believe the null hypothesis (we call this 'rejecting' the null hypothesis). So what we are trying to do is decide whether or not to reject the null hypothesis.

We concentrate on the null hypothesis because it is easier to prove that something is untrue than to prove that it is true. To prove the hypothesis 'all cats have a tail' definitively, you would have to check every cat in the world! But as soon as you find just one cat without a tail you have proved that the statement is definitely false. This is why we either reject the null hypothesis or say there is no evidence to reject it and therefore accept it.

So, to summarise, a hypothesis is a specific and predictive statement about the possible answers that could result from our research question. It primarily aims to answer our research question in terms of whether our results are significant or not – effectively they provide a yes or no answer to our research question.

Significance testing: The confidence interval method

The process of examining the null hypothesis in the light of sample evidence is called a **significance test**. It is possible to do this using confidence intervals. This is not the method generally used by most people, but it may help you to understand what is going on more easily.

Steps in significance testing using confidence intervals

Step 1: Write down the hypotheses

State the null hypothesis (what you would have expected to happen) and the alternative hypothesis (what actually appears to be happening). This should be done in terms of a population mean.

Step 2: Find the statistic for the sample

Calculate the mean for your sample.

Step 3: Find a confidence interval

Find a 95% confidence interval for the sample mean.

Step 4: Draw a conclusion

Does your population value fall within this range? If *yes*, accept H_0; if *no*, reject H_0.

EXAMPLE: COMPANY STAFF IQ

Let's go through the four steps for the example about the IQs of a company's staff.

Write down the hypotheses:

$$H_0: \mu = 120 \text{ (the staff have a mean IQ of 120)}$$

$$H_A: \mu \neq 120 \text{ (the staff do not have a mean IQ of 120)}$$

Note that the hypotheses can be written either in statistical jargon or in words.

Find the statistic for the sample: The mean IQ in our sample is 114. The sample mean is usually written $\bar{x}$, so we would write:

$$\bar{x} = 114$$

Calculate a range from the sampling distribution. Suppose we are prepared to be wrong one time in 20 (wrong 5% of the time and right 95% of the time). Then, for a normal distribution, we want to find the values that we are 95% certain that the population mean would lie between if the null hypothesis were true (if the company's staff did have a mean IQ of 120). This is shown in Figure 11.1.

This is just like calculating a confidence interval. The value 1.96 is used because we are testing at the 5% level; in other words, we are calculating a 95% confidence interval. Remember that the employees' original figures showed the standard deviation for their staff was 10. Therefore the **confidence interval** will be:

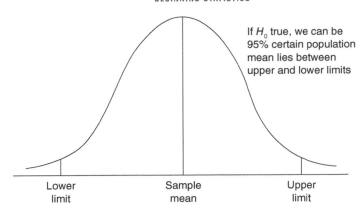

If H_0 true, we can be 95% certain population mean lies between upper and lower limits

Lower limit   Sample mean   Upper limit

Figure 11.1   The upper limit and lower limit

$$95\% \, CI = sample \, mean \pm 1.96 \times \frac{standard \, deviation}{\sqrt{n}}$$

$$= 114 \pm 1.96 \times \frac{10}{\sqrt{36}} = 114 \pm 3.27 = (110.73, 117.27)$$

Draw a conclusion: The population mean, 120, does not lie in the range (110.73, 117.27). The sample is very unlikely to have come from a population with a mean of 120. Therefore we can reject the null hypothesis H_0 and accept the alternative hypothesis H_A. We conclude that the staff in the company do not have a mean IQ of 120. The sample is significantly different from their claim, sufficient to make us believe that they were not telling the truth!

The test statistic method

The way we have just tackled this problem was a good introduction, but there is a much more commonly used method for testing hypotheses: the **test statistic** method.

Remember that we want the range where we accept H_0 to cover the middle 95% of the possible sample means. This corresponds to 47.5% either side of the population mean, as in Figure 11.2. Therefore the 5% of sample means outside this range will consist of 2.5% of sample means at either end of the distribution.

In terms of Z-scores, looking up 0.475 (the area 47.5%) in the middle of the normal table (Table A1.1) gives us a Z-score of 1.96 (check this yourself).

We want to reject the null hypothesis if the sample value is particularly extreme. So we can use 1.96 and −1.96 as the cut-off points. The points 1.96 and −1.96 are known as the **critical values**. If the Z-score in the sample lies in the **critical region** above 1.96 or below −1.96, we can reject H_0.

How can we find the Z-score for our sample? The formula is similar to the Z-score formula we learnt in Chapter 6, but the sample mean is compared to the population mean assumed under the null hypothesis, and because we are dealing with a sample, the standard error is used rather than the standard deviation:

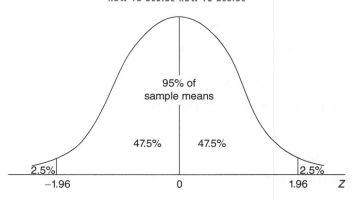

95% of
sample means

47.5%   47.5%

2.5%   2.5%

−1.96   0   1.96   Z

Figure 11.2   **Z-scores for the middle 95% of sample means**

$$Z = \frac{\text{sample mean} - \text{population mean}}{\text{standard error}} = \frac{\bar{x} - \mu}{\text{SD} / \sqrt{n}}$$

In this case, the Z-score will be:

$$Z = \frac{114 - 120}{10 / \sqrt{36}} = -3.6$$

The Z-score calculated like this is known as the test statistic. The test statistic −3.6 is much less than −1.96 (see Figure 11.3), so it definitely lies in the critical region. We reject the null hypothesis and conclude that the employees are extremely unlikely to have a mean IQ of 120. It is highly improbable that the sample comes from a population with a mean of 120 and a standard deviation of 10.

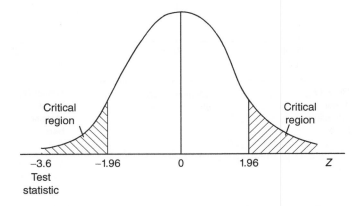

Critical   Critical
region   region

−3.6   −1.96   0   1.96   Z
Test
statistic

Figure 11.3   **A Z-score of −3.6 lies in the critical region**

Steps for a hypothesis test: Test statistic method

As a summary, here are the four steps needed for carrying out a hypothesis test comparing a sample mean to a population mean. Each step is important. If you forget to write down hypotheses at the beginning, the results will be meaningless.

Step 1: Write down the hypotheses

Write down the null and alternative hypotheses.

Step 2: Find the sample statistic

Find the sample mean. You may be told this or may need to calculate the mean yourself from some data.

Step 3: Calculate the test statistic

Calculate the test statistic Z:

$$Z = \frac{\bar{x} - \mu}{SD / \sqrt{n}}$$

Step 4: Compare the test statistic with the critical values

Compare the test statistic with the critical values, as in Figure 11.4. Draw a conclusion about the hypotheses and explain in words what this means.

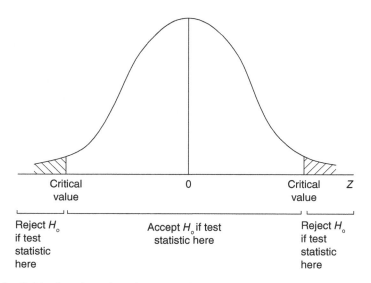

Figure 11.4   **Critical regions for a hypothesis test**

EXAMPLE: CLASS SIZES

There has been much research conducted about the potential impact of class size on educational attainment. Therefore, it is an issue parents are particularly concerned about. But how can it be determined whether class sizes are significantly different from national averages?

A group of parents believe that class sizes in their local primary schools are too high. They find that in a sample of 22 classes in the town, the mean number of children per class is 33. Let's suppose that, according to national education figures, class sizes nationally are normally distributed with a mean of 30 children and a standard deviation of 8 children. Are class sizes in the town significantly different from class sizes nationally?

Let's go through the four steps:

Write down the hypotheses:

H_0: μ = 30 (class sizes in the town are no different to the national average)

H_A: $\mu \neq$ 30 (class sizes in the town are different from the national average)

Find the sample statistic: we have been told that the mean class size in the town is 33.

Calculate the test statistic:

$$Z = \frac{33 - 30}{8 / \sqrt{22}} = 1.76$$

Compare the test statistic with the critical values: the test statistic 1.76 lies between −1.96 and 1.96, as shown in Figure 11.5. Therefore we do not have sufficient evidence to reject H_0. We must accept the null hypothesis that class sizes in the town are not significantly different from class sizes nationally.

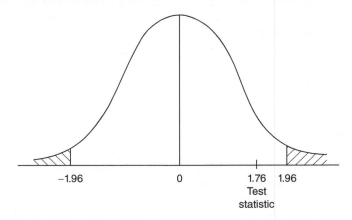

Figure 11.5   A *Z*-score of 1.76 does not lie in the critical region

Important note

In both these examples, we knew the standard deviation of the population. Often, we will not know the standard deviation of the population, but can instead calculate the standard deviation for the sample. The sample standard deviation can then be used to calculate the test statistic but, in general, this should only be done if the sample size is at least 30. Otherwise a different test must be used (see Chapter 12).

Tests for proportions

So far, we have looked at hypothesis tests for a mean, where we are comparing the mean of a sample with that specified by a hypothesis. Sometimes, however, we may have a sample proportion which we want to compare to a population proportion. The method is the same as that for a Z test using means, except for the calculation of the test statistic.

The following formula gives the calculation of a Z-statistic to compare a sample proportion p with the population proportion Π:

$$Z = \frac{p - \Pi}{\sqrt{\Pi\,(1 - \Pi)/n}}$$

The standard error for a proportion, rather than the standard error for a mean, is used on the bottom line. Note that we always use the hypothesised population proportion Π to calculate the standard error.

EXAMPLE: RACIAL DISCRIMINATION

A London clothing store claims that it does not racially discriminate. Half of its employees come from ethnic minorities and half do not. You observe that 23 out of 28 people who were fired last year came from an ethnic minority. Is it likely that the store is discriminating against such employees?

State the hypotheses:

H_0: $\Pi = 0.5$ (the proportion of those fired coming from ethnic minorities is the same as the proportion in the company's workforce, i.e. 0.5)

H_A: $\Pi \neq 0.5$ (the proportion of those fired coming from ethnic minorities is different from 0.5)

Find the sample statistic: The proportion fired who come from an ethnic minority will be 23 divided by 28:

$$p = \frac{23}{28} = 0.82$$

Calculate the test statistic:

$$Z = \frac{p - \Pi}{\sqrt{\Pi(1 - \Pi)/n}} = \frac{0.82 - 0.5}{\sqrt{0.5(1 - 0.5)/28}} = 3.39$$

Compare the test statistic with the critical values: for a Z test at the 5% level, the critical values are −1.96 and 1.96. The test statistic 3.39 is much greater than 1.96 and lies in the critical region. Therefore we should reject H_0 and conclude that there is evidence that the company is guilty of racial discrimination.

The level of significance

All the examples so far use tests at the 5% level of significance. But what if we want to be extremely cautious and test at the 1% level (we are only prepared to be wrong 1 time in 100)?

Steps 1–3 are the same. Only step 4 changes: we use different critical values. The critical region now only covers 1% of the area rather than 5%. In a two-tailed (two-sided) test, the 1% region is split between the two tails with 0.5% on each side, as shown in Figure 11.6. Therefore the percentage of the total area between the mean and a critical value will be 49.5%. Looking up the area 0.495 in the middle of the normal table (Table A1.1) gives a Z-score of 2.575. So the critical values for a two-tailed test at the 1% level are −2.575 and 2.575.

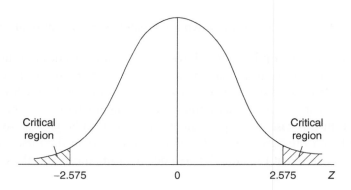

Figure 11.6 Critical regions for a two-sided test at the 1% level

In this chapter's first example, about the IQ of company staff, the test statistic was −3.6, so we would still reject H_0 at the 1% level. In the second example, regarding class sizes, H_0 was not rejected at the 5% level so it cannot possibly be rejected at the 1% level. In some cases, you may reject H_0 at the 5% level but not reject it at the 1% level (if the test statistic lies between 1.96 and 2.575, or between −2.575 and −1.96, for a two-sided test). You are more likely to reject H_0 at the 5% level than the 1% level because the rejection region is larger.

Note that you may want to memorise the critical values for each type of test. However, if you know how to work out the Z-score from the area as shown in the diagrams, you should be able to work out the critical values for a one- or two-tailed test at any level of significance, providing you have the normal tables with you.

One-sided tests

The tests we have carried out so far have been two-sided tests, otherwise known as two-tailed tests. In a two-sided test, the aim is to discover whether the sample mean or proportion is *different* from the population mean or proportion, not whether it is either larger or smaller.

Sometimes we are fairly sure that the sample mean is, for example, larger than the population mean. In other words, we know in which direction the difference is. In the example about class sizes, the parents were fairly sure that the class sizes in their town were *bigger* than the national average. So instead of testing to see whether their class sizes were *different* from the national average, we could test to see whether they are bigger.

To do this we use a one-sided (or one-tailed) test which predicts the direction of a difference or relationship. The procedure is similar, apart from the hypotheses and the critical values.

For a one-sided test, the null hypothesis is the same as for a two-sided test, but the alternative hypothesis becomes directional. The new hypotheses might be:

H_0: $\mu = 30$ (class sizes in the town are the same as the national average)

H_A: $\mu > 30$ (class sizes in the town are greater than the national average)

The test statistic will still be 1.76. However, the critical values are different. In a two-sided test, the 5% critical region was split into the two tails of the distribution. With a one-sided test, the 5% critical region is all at one end; which end depends on whether you are testing for a sample mean being greater or less than the population mean. In this case, we expect the sample mean to be greater, so the critical region will be on the right-hand side, which is higher than the mean, as shown in Figure 11.7. With 5% at the end, the area between the mean and the critical value

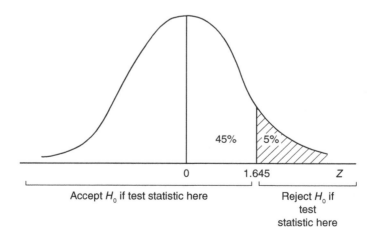

Figure 11.7 Critical region for a one-sided test (where the sample statistic is expected to be bigger than the population statistic)

will be 45%. If you look up the area 0.45 in the middle of the normal table (Table A1.1), this gives a Z-score of 1.645. (Check this yourself: in fact you will see that 1.645 is the mean of the two nearest Z-scores.)

So for a one-sided test, the critical value at the 5% level is 1.645, or −1.645 if you are testing for the sample mean being less than the population mean.

In this example, the test statistic of 1.76 lies in the critical region, so we can reject the null hypothesis and conclude that the class sizes in the town are significantly bigger than the national average.

This is a different conclusion than the one reached using a two-tailed test. It is always easier to reject the null hypothesis using a one-sided test, because the Z-score does not need to be as large for it to lie in the critical region. The null hypothesis can be rejected on weaker evidence and so one-tailed tests should be used with caution. It is essential that you use them only when you are absolutely certain of the direction of the effect. You should always remember that if you specify a one-tailed hypothesis then you do not have the option of moving to a two-tailed hypothesis if your sample turns out to have a mean or proportion which has a value in the opposite direction.

EXAMPLE: STUDENT FEES

A student union claims that 95% of their students disagreed with the rise in tuition fees. A student wants to explore this topic further for their dissertation. They take a random sample of 40 students at the same university and find that 33 of them disagreed with the rise in tuition fees, with the other seven feeling that the rises were a necessary measure, given the state of the economy and the increasing opportunities in the future it was likely to provide. The student researcher felt that this may be a more appropriate figure than that presented by the student union. Has the student got a good case?

The student is convinced that a *smaller* proportion of students disagree with the fee increases than the student union claims. We therefore use a one-sided test.

Hypotheses:

$$H_0: \Pi = 0.95$$

$$H_A: \Pi < 0.95$$

Sample proportion:

$$p = \frac{33}{40} = 0.825$$

Test statistic:

$$Z = \frac{0.825 - 0.95}{\sqrt{0.95(1 - 0.95)/40}} = -3.63$$

Because we are testing to see whether the proportion of students who disagreed with the rise in tuition fees is *smaller* than 0.95, the critical region will be below the mean, as in Figure 11.8. The critical value will be –1.645. As –3.63 is smaller than –1.645, it lies in the critical region. We can reject the null hypothesis and conclude that there is evidence that, based on this sample, the proportion of students who disagree with fee increases is significantly lower than 0.95.

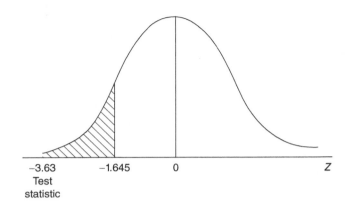

Figure 11.8 A *Z*-score of –3.63 lies in the critical region for a one-sided test

Note that if we had used a two-sided test for this example, the null hypothesis would still have been rejected because –3.63 is less than the –1.96 critical value used in a two-sided test at the 5% level.

Summary

This chapter has introduced you to the use of hypotheses, designed to express relationships between variables and, crucially, whether there is a significant relationship between those variables. In addition, they are important in relation to testing how a particular sample of a variable compares with the population. It has shown how to examine the null hypothesis in the light of sample evidence in the form of significance testing using confidence intervals, and how to use the test statistic method to test hypotheses using Z-scores when you have sample means or proportions. It has also covered the difference between two-sided (or two-tailed) tests and a one-sided (or one-tailed) test. In the next chapter we will move on to look at using one-sample *t* tests to compare the sample mean with the population mean when the sample is small, and one-sample sign tests to compare the median of the sample with the population median.

PRACTICE QUESTIONS

11.1. You are interested in the drinking habits of school children and carry out a survey of 34 girls aged 14 from a local school who had drunk in the last week. The mean number of units of alcohol consumed by the girls in the previous week was 10 units with a standard deviation of 0.8 units. National data suggest that mean weekly alcohol consumption among girls aged 14 who had drunk in the last week is approximately 8.2 units per week (Fuller, 2013).

Carry out a test at the 5% level to determine whether alcohol consumption among girls in the sample is significantly different from the national average.

11.2. A group of 11 professional social workers has a mean basic annual salary of £26,555. However the Local Government Earnings Survey 2010/11 states that the mean annual basic salary of full-time social workers is £30,386 (Local Government Group, 2011). Let's suppose that the standard deviation was £6,321.

The social workers complain to the higher authorities that they are badly off. Carry out a one-sided test at the 5% level to see whether their claim is justified.

11.3. When a politician was elected onto the city council, 63.8% of voters thought that local crime prevention was an important issue. Since then, several measures have been taken to reduce crime and the politician wonders whether attitudes have changed. In a recent survey of 350 voters, 204 still thought that local crime prevention was an important issue. Perform a hypothesis test at the 5% level to determine whether there has been a significant change in this attitude over time.

11.4. The mean hourly rate of pay for white males in full-time employment was £12.94 in 2007/8 according to Labour Force Survey data (Longhi and Platt, 2008). A researcher is studying people of Pakistani origin in Britain and finds that the comparable rate of pay in her sample of 55 men is £9.99 with a standard deviation of £2.12.

Carry out a two-sided test to see whether the difference is significant:

(a)   at the 5% level;
(b)   at the 1% level.

11.5. A researcher claims that anybody conducting a small-scale postal survey on an issue of local concern will have a response rate of 40%. You recently tried to do a small-scale postal survey about the need for local sports facilities and only had 18 replies from a sample of 60.

Does your experience provide any evidence at the 5% level to suggest that the researcher's claim might be incorrect?

11.6. A local health centre asks all smokers registered with the practice whether they would like to give up smoking. Of 124 smokers, 89 say that they would like to give up.

A national survey carried out in 2010/11, the General Lifestyle Survey (Office for National Statistics, 2013e), suggested that 63% of adult smokers would like to give up smoking. Is the proportion of smokers at the health centre who want to give up significantly different from the national proportion?

TWELVE

MORE TRICKY DECISIONS

Introduction

So far we have carried out Z tests comparing a sample to a population. This can only be done if we know the standard deviation of the population, or if we don't then the sample size must be 30 or greater. It should be noted that, in our experience, most social scientists use data which satisfy these assumptions. If we do not know the population standard deviation and the size of the sample is below 30, a slightly different approach must be used. This chapter will look at two methods for testing small samples where the population standard deviation is unknown.

- The one-sample *t* **test** compares the sample mean with the population mean, just like a Z test. It is very similar, but the critical t table is used instead of the normal table. The t test can only be used when the data are normally distributed.

- The one-sample **sign test** compares the median of the sample with the population median. This test can be used when the data do not follow the normal distribution; in other words, it is suitable for skewed data. It can also be used if we want to compare medians rather than means.

Time will be spent working through examples which use these new tests. By the end of the chapter you should be able to:

- Identify when it is appropriate to use different kinds of statistical tests, specifically Z-scores, t tests and one-sample sign tests.
- Understand how to calculate significance levels by using t tests and one-sample sign tests.
- Understand how the distribution of the data influences the level of central tendency, and the statistical test to use.

Deciding which test to use

Figure 12.1 may help you to decide which is the most appropriate test to use for a particular dataset. When presented with a problem, always write down what you know first, as this will help you decide which test to use. For example, if you realise that you do not know the population standard deviation, but there are 40 people in the sample, you can use a Z test. If you find that you do not know the population standard deviation and there are fewer than 30 people in the sample, drawing a stem and leaf plot can help you to decide whether the data are normally distributed or not. Because we are dealing with small samples, it is harder to reject the null hypothesis using these tests than with a conventional Z test. Smaller samples are less reliable and so we must be more certain of a difference before rejecting H_0. Where possible, the better strategy is still to obtain a larger sample to work from.

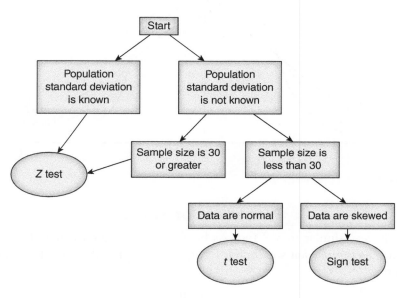

Figure 12.1 How to decide which test to use

The *t* test

A t test is only used when the population standard deviation is unknown, the sample size is less than 30 and the sample data are normally distributed. A t test is very similar to a Z test, but rather than using the normal distribution, the **t distribution** must be used instead, to take account of the small sample size. There are three different types of t test. When the same sample of respondents is measured at two different times or two carefully matched samples are measured against each other, this is called a paired-sample t test. A matched t test like this is used in situations where two measurements are taken for each respondent and is often used in experiments where there are before-treatment and after-treatment measurements. The process of exploring two samples of respondents which are unmatched and separate from each other is known as the independent-sample t test. Finally, the one-sample t test is where one sample of respondents whose results are known is compared with some known national figures, for instance. It is the one-sample t test that this chapter concentrates on (further information about other types of t test can be found in other statistics books).

The *t* distribution

The *t* distribution is similar to the normal distribution, but there is a separate *t* curve for each sample size, as shown in Figure 12.2. As the sample size increases, the *t* curves get taller and thinner. Once the sample size reaches 30 or more, the *t* curve is approximately the same as the normal distribution. This is why we can use the normal distribution as an approximation where $n \geq 30$.

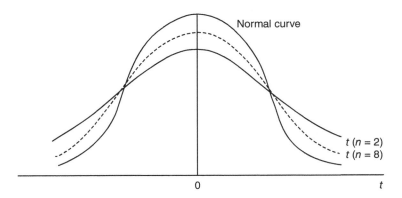

Figure 12.2 The *t* distribution

Procedure for a *t* test

A *t* test is very similar to a *Z* test. As before, the hypotheses should be written down and the sample statistic found. The test statistic has the same formula as for the *Z* test:

$$t = \frac{\overline{x} - \mu}{SD / \sqrt{n}}$$

The main difference lies in finding the critical values. These are found using the critical *t* table (see Table A1.2 in Appendix 1). You will see that the first column in the *t* table is called 'df' which stands for 'degrees of freedom'. In a one-sample test such as those shown here, the number of degrees of freedom is always one less than the number in the sample.

The theory behind degrees of freedom is that we use the sample values to estimate the population mean (or proportion) and the variability. It is the number of sources of variation in the data examined. In a one-sample test where you have, for example, 20 observations in your sample, estimating the population mean takes up one observation so the remaining 19 observations are free for estimating the variability. So the *t* curve to use is the curve for $n - 1$ degrees of freedom. For example, if the sample size is 15, t_{n-1} is t_{14}. To find the value of *t* for 14 degrees of freedom, find 14 in the first column (df) and look horizontally along.

For a two-tailed test at the 5% (or 0.05) level, look down the $t_{df}(0.025)$ column. This is because for such a test there will be 95% in the middle of the distribution and 2.5% (0.025) at each end, as shown in Figure 12.3. So for a 5% level test with

14 degrees of freedom, the *t* value from the table is 2.1448 and the critical values will be –2.1448 and 2.1448.

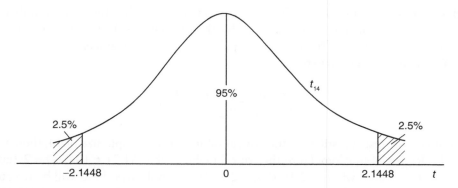

Figure 12.3 Critical regions for a two-sided test at the 5% level with 14 degrees of freedom

Similarly, for a two-tailed test at the 1% level, look down the $t_{df}(0.005)$ column, because there will be 99% in the middle of the distribution and 0.5% (0.005) in each tail. Can you work out which columns you would use for one-tailed tests at the 5% and 1% levels? (Sketch a diagram to help.)

EXAMPLE: DRINKING HABITS

The General Lifestyle Survey 2010 showed that the mean weekly alcohol consumption among women aged 16–24 was 8.4 units (Dunstan, 2012). You wonder whether your fellow students are different in this respect and undertake a survey of drinking in the previous week. The number of units consumed by a sample of female students in the previous week is shown in Table 12.1.

Table 12.1 Units of alcohol consumed by 14 female students in the previous week

11	0	12	23	17	6	1
17	11	14	0	18	14	5

To carry out a hypothesis test at the 5% level, let's follow the key steps as before. A *t* test must be used because we do not know the population standard deviation and the sample size is less than 30.

Write down the hypotheses:

H_0: μ = 8.4 (alcohol consumption among female students is the same as the national mean for women aged 16–24)

H_A: $\mu \neq 8.4$ (alcohol consumption among female students is different from the national mean for women aged 16–24)

Find sample statistics: From the data, we can calculate that the mean number of units consumed in the previous week by the sample was 10.64 with a standard deviation of 7.26 units. (Check these figures yourself for practice.)

Calculate the test statistic:

$$t = \frac{\bar{x} - \mu}{SD / \sqrt{n}} = \frac{10.64 - 8.4}{7.26 / \sqrt{14}} = 1.15$$

Compare the test statistic with the critical value: With a sample size of 14, there will be 13 degrees of freedom. From the critical t table (Table A1.2), $t_{13}(0.025) = 2.1604$. The critical values will be -2.1604 and 2.1604. The null hypothesis will be rejected if the test statistic is less than -2.1604 or greater than 2.1604 (see Figure 12.4). The test statistic 1.15 lies between -2.1604 and 2.1604 and so the null hypothesis is accepted. From this sample, there is no evidence that alcohol consumption among female students is different from national levels for females aged 16–24.

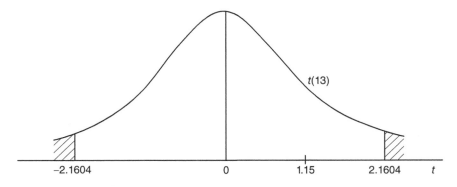

Figure 12.4   A t value of 1.15 does not lie in the t_{13} critical region

---

EXAMPLE: PSYCHOMETRIC TESTING

A group of social science graduates applying for management jobs in a company are required to take psychometric tests. Their scores for the test are shown in Table 12.2. The mean score for the test among all applicants is 62. Explore whether there is any evidence that the social science graduates are better than the mean at the psychometric tests?

Table 12.2   Psychometric test scores for 12 social science graduates applying for management jobs in a company

71	63	62	74	69	67	59	65	68	65	66	67

The hypotheses could be written as follows:

H_0: μ = 62 (the performance of the social science graduates is the same as the mean)

H_A: μ > 62 (the performance of the social science graduates is better than the mean)

Note that this is a one-sided test.

The sample statistics must be calculated from the data. The mean score from the sample of students is 66.33 with a standard deviation of 4.03.

Next the test statistic is calculated:

$$t = \frac{\bar{x} - \mu}{SD / \sqrt{n}} = \frac{66.33 - 62}{4.03 / \sqrt{12}} = 3.72$$

Finally, the critical values must be found and a conclusion drawn. With a sample size of 12, there will be 11 degrees of freedom. For a one-sided test at the 5% level, there will be 5% of the area in one tail, so the $t(0.05)$ column should be used. From Table A1.2, $t_{11}(0.05) = 1.7959$, so the critical value will be 1.7959. Figure 12.5 illustrates this process.

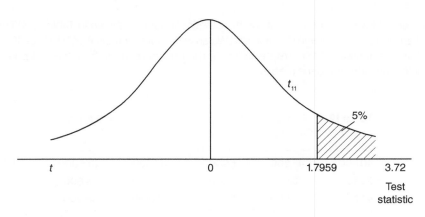

Figure 12.5 A t value of 3.72 lies in the t_{11} critical region

The test statistic 3.72 lies in the critical region and so the null hypothesis can be rejected. There is evidence that the sample of social science graduates has done significantly better than the mean in the psychometric tests.

The sign test

If the distribution of a dataset is skewed, the median is a better measure of the average than the mean because it is not affected by extreme values (you should remember this from Chapter 4). The sign test compares the median of a sample with a population median and is used with small samples of skewed data. It can also be used when the population median is known rather than the population mean.

Idea behind the sign test

The mean has a sampling distribution, as described by the central limit theorem, which helps us to compare a sample mean with a population mean. However, the median does not have a similar sampling distribution. (This is why the sign test is known as a **non-parametric test**: 'non-parametric' tells us that there is no sampling distribution.)

All we know is that the median lies in the middle of the distribution. In the population, we know that half of the observations will lie above the median and half below it. If there is no evidence that the sample is no different than the population, we would expect that about half of the observations in the sample will lie above the population median and half below it. If the sample is genuinely different from the population, the proportion of observations above the median will be markedly greater or smaller than 0.5.

EXAMPLE: THE COST OF PRISONS

The cost of keeping prisoners at the hypothetical prison is shown in Table 12.3. The median cost per prisoner in England and Wales was approximately £37,000 in 2010–11 (Ministry of Justice, 2012). Are the costs of the prisoners in the hypothetical prison significantly different to the median?

Table 12.3 Costs of keeping prisoners in the hypothetical prison in 2010–11 (£)

40,500	45,200	55,500	33,900	37,500
35,400	35,500	47,000	44,900	34,800
44,700	38,900	25,500	40,600	38,000

The hypotheses are always written in terms of the proportion of observations above the median. Here Π_m is the proportion of observations above the median in the population and p_m is the proportion in the sample. So we can write:

H_0: $\Pi_m = 0.5$ (the cost of keeping prisoners in the hypothetical prison comes from a population with half the observations above £37,000)

H_A: $\Pi_m \neq 0.5$ (the cost of keeping prisoners in the hypothetical prison does not come from a population with half the observations above £37,000)

To find the test statistic p_m, the proportion of prisoner costs above the median in the sample must first be found in the hypothetical prison. To do this, put a plus or minus sign by each observation to indicate whether it is above (+) or below (−) the median:

40,500 +	45,200 +	55,500 +	33,900 −	37,500 +
35,400 −	35,500 −	47,000 +	44,900 +	34,800 −
44,700 +	38,900 +	25,500 −	40,600 +	38,000 +

The proportion above the median will be the number of plus signs divided by the total number in the sample:

$$p_m = \frac{\text{number of +signs}}{\text{total number}} = \frac{10}{15} = 0.67$$

(If you have one or more observations equal to the median, take half of them to be above and half below.)

A Z test statistic can then be found in the same way as for a hypothesis test for a proportion. If the null hypothesis is true, the population proportion above the median will be 0.5 and the standard error of the population will be

$$\sqrt{\frac{0.5(1-0.5)}{15}}$$

The Z test statistic in this case would be:

$$Z = \frac{p_m - \Pi_m}{\sqrt{\Pi_m(1-\Pi_m)/n}} = \frac{0.67 - 0.5}{\sqrt{0.5(1-0.5)/15}} = 1.32$$

For a two-tailed test at the 5% level, the critical values will be −1.96 and 1.96. The test statistic 1.32 lies between −1.96 and 1.96, so we have no reason to reject the null hypothesis. We must accept the null hypothesis that the cost of the keeping prisoners in the hypothetical prison is not significantly different from the median.

Even though it appeared that the costs of the keeping prisoners in the hypothetical prison in the sample were higher than the median, we cannot reject the null hypothesis. Non-parametric tests are less likely to find a significant difference between the sample and the population than tests based on the normal distribution (Z tests). Remember that the sign test does not use all the information from the data. It only takes account of whether each observation is above or below the median and not its actual value. Therefore we recommend this test is used only when the assumptions surrounding the use of the Z test are clearly not satisfied. Finally, it should be noted that there are a large number of non-parametric tests for use in different situations and you are advised to consult a more specialised book (there are many) on non-parametric tests if appropriate. More information about where to look is provided in Chapter 15.

EXAMPLE: BOOK PRICES

The median cost of textbooks in the local bookshop in 2012 was £24.99. In early 2013 you took a sample of 25 textbooks from the shop and found that 18 of them cost more than £24.99. Is there any evidence that the price of textbooks changed in 2013? (Use a two-tailed test at the 5% level.)

This problem is phrased slightly differently in that some of the work is done for us; we know that 18 out of 25 books in the sample cost more than the median price of the 2012 population. Otherwise, the method is the same, as the workings below show:

$$H_0: \Pi_m = 0.5$$

$$H_A: \Pi_m \neq 0.5$$

$$p_m = \frac{18}{25} = 0.72$$

$$Z = \frac{0.72 - 0.5}{\sqrt{0.5(1 - 0.5)/25}} = 2.2$$

For a two-tailed test at the 5% level, the critical values are −1.96 and 1.96. The test statistic 2.2 is greater than 1.96 and so we can reject the null hypothesis. There is evidence to suggest that the price of textbooks did increase in 2013, so you should go and complain! You will probably be told that the global economy is to blame and the poor bookseller can do nothing!

Moving on ...

This chapter has concentrated on one-sample tests. However, it is worth recognising that it is also possible to compare statistics from two samples. For instance, we might be interested in whether the number of times men had gone to the gym in the last few weeks is different from the number of times women have done. Here there are two independent samples (one of men and one of women) and it is possible to calculate a summary statistic. This allows us to explore whether any differences in the means from the two samples (of men and women) are because we are dealing with random samples that aren't representative, or whether the differences in the means are due to real differences in the populations from which they are drawn. While there are similarities with one-sample test procedures and the general logic of hypothesis testing remains, the process of calculating standard errors differs and is beyond the scope of this book.

Summary

It is apparent that the test that you should use depends on the assumptions it fulfils. This chapter has provided details about when Z tests, one-sample t tests and one-sample sign tests in particular should be used. For instance, you should now be aware that the one-sample t test compares the sample mean with the population mean, just like a Z test, but when the sample is small, and it can only be used when data are normally distributed. So it is used when you want to identify whether the value observed is significantly different from the mean value of a variable. The sign test, on the other hand, compares the median of a sample with a population median and is used with small samples of skewed data. It is useful when the population median is known rather than the population mean. You should now be able to calculate significance levels by using these tests to test a hypothesis.

PRACTICE QUESTIONS

12.1 Davis et al. (2012), using the Living Costs and Food Survey (previously known as the Expenditure and Food Survey), found that the mean weekly household expenditure among lone-mother families with at least one dependent child is £297. From interviewing 16 lone mothers in your local area, you find that they spend on average £274 per week with a standard deviation of £33. Is there any evidence at the 5% level that the household expenditure of lone mothers in your area is different from the national average?

12.2 A class of 25 reception children is asked how much pocket money they receive weekly. The mean amount received was found to be £1.56 with a standard deviation of £0.45.

You find data which suggest that the mean amount of pocket money received by children in the country is £1.80. Use a one-tailed test at the 5% level to determine whether the children are receiving less pocket money than average.

12.3 The mean weekly childcare cost for 25 hours of nursery care for a child under the age of 2 in London is £133.17 (Daycare Trust and Family and Parenting Institute, 2013). For the

Table 12.4 Average weekly childcare cost for 25 hours of nursery care for a child under age 2 by region, 2012–13

Region	Cost of childcare (£)
East of England	106.55
East Midlands	104.91
North East	101.61
North West	92.22
South East	125.16
South West	113.32
West Midlands	96.92
Yorkshire and Humberside	102.71

Source: Daycare Trust and Family and Parenting Institute, 2013

(Continued)

(Continued)

purposes of this exercise let's suppose this was also the median and that Table 12.4 shows the median costs in other regions of England. Use a sign test at the 5% level to determine whether the cost of childcare in London is higher than in other regions.

12.4 A charity runs an annual sponsored bike ride. Last year the median amount raised per rider was £63.20. This year the charity hopes that the riders have brought in more sponsorship money. To see whether this is the case, they take a sample of 20 riders and check how much money each raised. The results (to the nearest 10p) are shown in Table 12.5.

Carry out a sign test (5% level) to find out whether the amount of sponsorship money raised per rider has changed significantly since last year.

Table 12.5 Sponsorship money raised by 20 riders in a charity ride (£ to nearest 10p)

23.40	52.00	39.20	48.50
65.10	69.90	68.80	101.30
110.70	44.70	82.20	88.60
77.30	72.60	99.40	61.10
60.30	71.40	59.80	66.70

THIRTEEN

CORRELATION AND REGRESSION

Introduction

You may have found yourself wanting to compare two variables with each other, for example, to answer questions such as:

- Is the social deprivation level of a district related to the level of unemployment in the district?
- Is the number of units of alcohol a student drinks in an average week related to how often they miss the Quantitative Methods class in a term due to a hangover?

You may also want to predict the value of one variable from the other, for example:

- If a district has an unemployment rate of 13%, how socially deprived would we expect it to be?
- If student X drinks 25 units in an average week, how many classes is he/she likely to miss in the semester?

Fortunately, statistics help us to address these sorts of question. Two methods which will be considered in this chapter and which will assist you in answering these types of question are the **correlation** and **regression**. Correlation measures the association between two continuous variables (interval or ratio) – in other words, the strength of the relationship between the values of the two variables. To quantify an association (to measure its strength numerically) we can calculate the **correlation coefficient**, a process you will learn in this chapter. Regression takes things a little further and enables us to predict the values of one variable from the values of another. So if an ice cream sales person knows the weather forecast (and it is reliable!) they could use this to predict the number of ice cream sales (it makes me feel hungry thinking about this example!). In many ways you can think of correlation and regression as being descriptive statistics on two variables as opposed to one. They are really useful in the social sciences in trying to establish associations. This chapter will take you through the process of working with and calculating correlation coefficients and regression analysis, and understanding how to interpret them and present them graphically. By the end of the chapter you should be able to:

- Identify when it is appropriate to use correlation coefficients and regression analysis
- Understand how they are represented graphically in scatterplots and line graphs
- Work out correlation coefficients, carry out regression analysis and interpret your findings

Correlation

Correlation measures the association between two continuous variables. The word 'correlation' is made of 'co-', meaning 'together', and 'relation', which says it all really. Variables can have a positive or a negative relationship with each other. Correlation is positive when the values increase together and negative when one value decreases as the other increases. These patterns are easy to see when presented graphically, as you will see shortly.

Here are some examples of variables whose values are associated:

- The number of years of education that women in developing countries have and the number of children they have. With more years of education, women tend to have fewer children.
- The sales of Zimmer frames and the number of people over 70 years old in the population. In populations with a high proportion of over-70s, we would expect sales of Zimmer frames to be high.
- A rather obvious temperature example is the higher the temperature, the more people are likely to get sunburn.

In some cases, two variables may be closely associated, but that does not mean that there is a **causal** relationship between them. If the value of one variable increases as the other variable increases (or decreases), it does not necessarily mean that one variable explains the other or could be used to predict the other. This can be seen in the two following examples.

EXAMPLE: IQ AND MENARCHE

Early research found that a high IQ and early menarche (starting menstrual periods) were associated among females. However, it was clear that having a high IQ did not cause the menarche to be early; neither did an early menarche cause a high IQ! In fact a third factor, social class, was influencing both variables (see Figure 13.1). Girls of higher social class were better nourished and therefore experienced menarche earlier than girls of lower social class. Girls of higher social class also had higher IQs due, probably, to better education and opportunity in the home compared with those of a lower social class. Therefore IQ and age at menarche were only associated because of the confounding effect of social class.

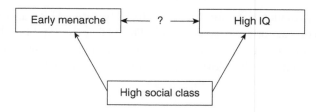

Figure 13.1   A causal link between early menarche and high IQ?

EXAMPLE: FOOTBALL TEAM PERFORMANCE AND ATTENDANCE

The chair of a football team achieving poor results assumes that this has caused decreased attendance figures. He is surprised, as in previous seasons when results have not been going well there has not been much difference in attendance figures. He decides to explore a fans forum to find out a bit more about fans' thoughts on the performances. While doing so, he discovers that a third variable may be having an effect on attendances: price rises put into effect recently. This leads him to explore special offers to encourage attendance. The other interesting issue with this example is whether results are affecting attendance or the attendance and lack of crowd is affecting performance. Therefore it is difficult to be sure about the direction of the possible relationship.

Scatter graphs

When you are trying to examine the association between two variables X and Y, always draw a **scatter graph** (also known as a scatterplot or scattergram) before doing anything else.

Table 13.1   Total fertility rate and contraceptive prevalence among women who are married or in a relationship in 10 countries

Country	Total fertility rate, 2010	% of women using contraception (any method), 2006–10
UK	1.9	84
USA	2.1	79
Congo	4.5	44
Egypt	2.7	60
Mexico	2.3	71
Brazil	1.8	80
Uganda	6.1	24
Turkey	2.1	73
Niger	7.1	12
Tanzania	5.5	34

Source: World Health Organization, 2012; United Nations, 2011

Table 13.1 provides data for the total fertility rate (TFR) and percentage of women using contraception in 10 countries. The TFR is the total number of children a woman would expect to have during her lifetime if current levels of childbearing prevailed. These data can be plotted on a scatter graph as in Figure 13.2.

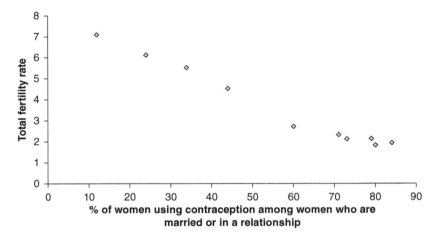

Figure 13.2 Total fertility rate and the percentage of women using contraception among women who are married or in a relationship in 10 countries

Source: World Health Organization, 2012 and United Nations, 2011

Tips for drawing scatter graphs

Remember all the rules for drawing graphs discussed in Chapter 3. In particular, take care in choosing the scale.

If you are interested in causation, as opposed simply to association, it is important to decide which of the two variables is the **dependent variable** and which is the **independent variable**. If you are trying to predict one variable from the other, the variable being predicted is the dependent variable. In this example, we know that the level of childbearing is likely to depend on the level of contraceptive use and so the TFR is the dependent variable and contraceptive prevalence is the independent variable. Often it will be obvious which variable is dependent, but sometimes it is not clear and in such cases it does not matter which is which.

Deciding which variable is dependent is important, because on a scatter graph the dependent variable should always go on the Y axis. The independent variable is therefore plotted on the X axis.

Figure 13.2 shows that, in general, countries where a high percentage of women are using contraception have low TFRs, while countries where contraceptive prevalence is low tend to have high TFRs. This makes substantive sense.

Occasionally, you may want to plot scores of variables that have been standardised (Z-scores). Because Z-scores are negative as well as positive, you will need a cross-shaped axis, as in Figure 13.3, which shows some standardised weights and heights (hypothetical data). The graph shows, in general, that the taller a person is, the heavier they are likely to be. We call this type of relationship a positive correlation.

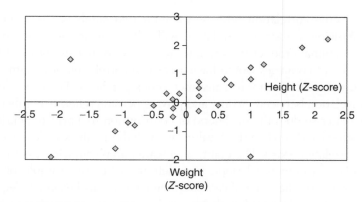

Figure 13.3 Standardised weights and heights of 25 people

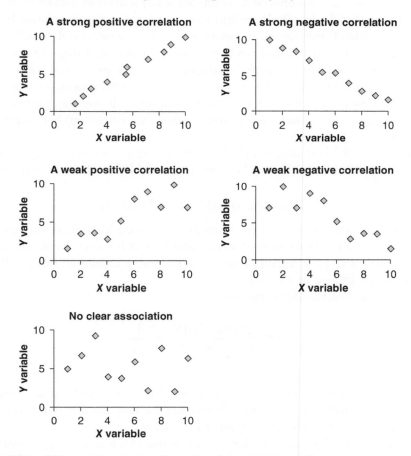

Figure 13.4 Different types of pattern found on scatter graphs

Interpreting scatter graphs

Figure 13.4 shows five examples of patterns which a scatter graph might show.

If when the values of the X variable are high the values of the Y variable are also high, and when the values of the X variable are low the values of the Y variable are also low, a positive correlation exists between the two variables. The points on a scatter graph showing a positive correlation will follow an upward slope from left to right. Gross national product (GNP) and the number of televisions per 1,000 population in different countries would have a positive correlation, for example, because richer countries will have more consumer goods per head. There would also be a positive correlation between income and private pension scheme membership, for as income increases so does the likelihood of contribution to a private pension.

On the other hand, if when the X values are high the Y values are low and vice versa, a negative correlation exists. The points on a scatter graph showing a negative correlation will follow a downward slope from left to right. An example of a negative correlation would be the value of cars and their ages: as cars become older, their value drops. Another example would be that beyond pension age, as your age increases, your self-reported health (measured on a continuous scale) is likely to decline.

If all the points lie in a fairly straight line, we have a **strong correlation** between the two variables. If they are more spread out, we have **weak correlation**. Therefore a correlation can be strong and positive, weak and positive, strong and negative or weak and negative! Note that the distinction between strong and weak is fairly arbitrary. You might want to describe a correlation as fairly strong, for example.

Finally, if the points appear to be randomly scattered all over the graph and show no clear pattern, we say that there is *no association* (no correlation) between the two variables.

You may be thinking that this is all a bit vague: ideally we want to quantify *how* weak or *how* strong a correlation between two variables is. This can be done!

The correlation coefficient

To quantify an association – in other words, to measure its strength numerically – we can calculate the **Pearson product moment correlation coefficient**, named after a statistician called Karl Pearson. You will be pleased to know that this is called the correlation coefficient or *r* for short.

Take a deep breath: here is the formula!

$$r = \frac{\Sigma[(X_i - \bar{X})(Y_i - \bar{Y})]}{\sqrt{\Sigma(X_i - \bar{X})^2 \, \Sigma(Y_i - \bar{Y})^2}}$$

The formula is honestly not as bad as it looks, although it can be time-consuming to work out by hand. After all, you should now be very familiar with expressions like $\Sigma(Y_i - \bar{Y})^2$ for calculating a standard deviation.

The way to handle a scary-looking equation like this is to draw up a worksheet to do each bit separately before finally putting them together in the equation.

Steps in calculating the correlation coefficient

The following tabulation calculates the correlation coefficient for the data on TFR and contraceptive prevalence in 10 countries (from Table 13.1). The procedure is as follows.

Step 1: Write down a list of all the observations for the two variables X and Y in the first two columns (A and B). Remember that X is contraceptive prevalence and Y is the TFR.

Step 2: For each variable, add up all the values of the observations and divide by the total number of them to obtain the means $\bar{X}$ and $\bar{Y}$ at the bottom of each column.

Step 3: In columns C and D, calculate the residuals $(X_i - \bar{X})$ and $(Y_i - \bar{Y})$ by subtracting the mean from each value.

Step 4: In columns E and F, square the answers that you got in columns C and D to obtain $(X_i - \bar{X})^2$ and $(Y_i - \bar{Y})^2$.

Step 5: In column G, multiply the values in column C by the values in column D to obtain $(X_i - \bar{X})(Y_i - \bar{Y})$.

Step 6: For columns E, F and G add up all the values in each column to obtain a total at the bottom of each column. These are the numbers that will go into the equation.

Note that a common mistake is to miss out column G and get an answer of zero for the top line of the equation, through thinking that the sum of the X residuals is zero and the sum of the Y residuals is zero, and so zero times zero equals zero. In fact you need to multiply *each* X residual by the corresponding Y residual and *then* sum these answers (as in column G). Your answer is unlikely to be zero.

A	B	C	D	E	F	G
X_i	Y_i	$(X_i - \bar{X})$	$(Y_i - \bar{Y})$	$(X_i - \bar{X})^2$	$(Y_i - \bar{Y})^2$	$(X_i - \bar{X})(Y_i - \bar{Y})$
84	1.9	27.9	−1.71	778.41	2.92	−47.71
79	2.1	22.9	−1.51	524.41	2.28	−34.58
44	4.5	−12.1	0.89	146.41	0.79	−10.77
60	2.7	3.9	−0.91	15.21	0.83	−3.55
71	2.3	14.9	−1.31	222.01	1.72	−19.52
80	1.8	23.9	−1.81	571.21	3.28	−43.26
24	6.1	−32.1	2.49	1030.41	6.20	−79.92
73	2.1	16.9	−1.51	285.61	2.28	−25.52
12	7.1	−44.1	3.49	1944.81	12.18	−153.91
34	5.5	−22.1	1.89	488.41	3.57	−41.77
$\bar{X} =$	$\bar{Y} =$		Totals	6006.90	36.05	−460.51
56.1	3.61					
				$\Sigma(X_i - \bar{X})^2$	$\Sigma(Y_i - \bar{Y})^2$	$\Sigma\left[(X_i - \bar{X})(Y_i - \bar{Y})\right]$

Now you are ready to put all these figures into the equation, which has magically been reduced to three numbers:

$$= \frac{-460.51}{\sqrt{6006.90 \times 36.05}} = -0.99 \text{ (to two decimal places)}$$

Eureka! There is just one problem here: what does an r value of −0.99 actually mean?

187

A brief guide to *r*, the universe and everything

The correlation coefficient *r* takes values between +1 and −1. This is what the different values of *r* indicate:

- *r* = +1: perfect positive correlation (all the points lie in a straight line)
- *r* between 0 and 1: positive but not perfect correlation
- *r* = 0: no association between the two variables
- *r* between −1 and 0: negative but not perfect correlation
- *r* = −1: perfect negative correlation (all the points lie in a straight line)

The nearer *r* is to +1 or −1, the stronger the relationship between the two variables. Values near zero indicate a very weak relationship.

In the example, the result *r* = −0.99 indicates a very strong negative correlation between contraceptive prevalence and the TFR (as we would expect from the scatter graph).

In other words, countries with a low percentage of women using contraception tend to have high TFRs and vice versa. Because the relationship is so strong, a country's TFR could safely be predicted from its contraceptive prevalence.

Note that in the social sciences, we would almost never get a perfect correlation between two variables because people never behave in exactly the way we would expect! For instance, richer people aren't always happier! A value as close to −1 as the example is quite rare.

Warning

The correlation coefficient measures **linear** association: how close the data points are to a straight line. However, with some relationships, the data follow a curved pattern. This is called a **curvilinear** relationship, as in Figure 13.5. In such cases *r* is *not* a good measure of the strength of association between the two variables. This is one reason why it is important to draw a scatter graph before doing any calculations: if it shows a curvilinear association, you should not calculate the correlation coefficient because it will be misleading.

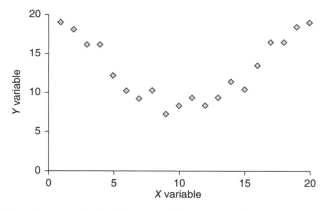

Figure 13.5 A scatter graph showing a curvilinear association

Regression: Predicting *Y* from *X*

When describing just one variable, we can summarise the data using means, medians and so on. But with two variables, we summarise their relationship using the equation of a straight line.

If *X* and *Y* are the two variables, the equation of a straight line is:

$$Y = a + bX$$

where *a* is the *Y* intercept and *b* is the gradient (explained in a moment).

(You may or may not remember from your school days that the equation of a straight line was $Y = mX + c$. This means exactly the same thing – we just use slightly different letters in statistics, so *a* is equivalent to *c*, and *b* is equivalent to *m*. If this means nothing to you, then just ignore it!)

If we calculate the values of *a* and *b*, this equation can be used to predict the value of the *Y* variable from any given value of the *X* variable. The problem is, how can we fit the best straight line to a set of data? With just two points, as in Figure 13.6, it's easy! But with three or more points, there are lots of possibilities.

Figure 13.7 shows some possible lines for a dataset with only three points. In a larger dataset with a strong correlation, it is usually possible to judge by eye roughly where the best fit line might go, but even so we would probably all draw it slightly differently. If the correlation is weak and there is a lot of scatter, it is very difficult to judge where the line should be. So we have to calculate the line of best fit, known as the regression line.

The line of best fit will be the line that minimises the vertical distance between the line and all the points. In jargon, the best fit line will minimise Σdi^2, where the d_i are the vertical distances between the line and each point as shown in Figure 13.8.

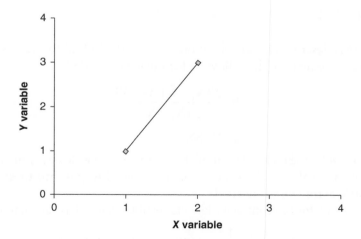

Figure 13.6 Finding the best straight line through two points

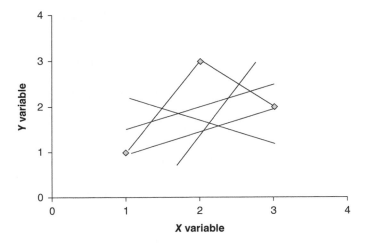

Figure 13.7 Finding the best straight line through three points

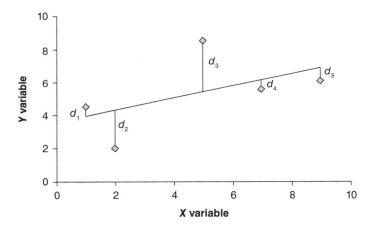

Figure 13.8 Distances between the line and the points

This line is described by the equation $Y = a + bX$. To find the values of a and b to put into the equation, the following formulae are needed:

$$b = \frac{\Sigma(X_i - \bar{X})(Y_i - \bar{Y})}{\Sigma(X_i - \bar{X})^2}$$

$$a = \bar{Y} - b\bar{X}$$

If you look back at the calculation of the correlation coefficient, you will see that we have already calculated all the numbers we need to put into these equations! So very little extra work is needed.

The value of b for the contraception and fertility example is calculated as follows:

$$b = \frac{-460.51}{6006.90} = -0.08 \, (\text{to two decimal places})$$

190

This value for b, along with the two means $\bar{X}$ and $\bar{Y}$, is used in the equation

$$a = \bar{Y} - b\bar{X} = 3.61 - (-0.08 \times 56.1) = 8.10$$

We now know the values of a and b. These are known as the regression coefficients. The equation of a straight line is $Y = a + bX$, so the equation of the regression line for these data will be:

$$Y = 8.10 + (-0.08X)$$

So the equation can be written as:

$$Y = 8.10 - 0.08X$$

The X variable represents the percentage using contraception and the Y variable represents the total fertility rate, so the variable names can also be put into the equation:

$$\text{TFR} = 8.10 - 0.08 \times \text{percentage using contraception}$$

And there we have it! But what on earth does it mean?

Interpreting the regression coefficients *a* and *b*

The regression coefficients a and b must always be interpreted in the *context* of the analysis that you are doing.

How should a be interpreted? In this case, $a = 8.10$. Looking at the equation TFR $= 8.10 - 0.08 \times$ percentage using contraception, we can see that if the percentage using contraception were zero, then -0.08×0 would equal 0 and so the TFR would take the value 8.10. Therefore we can **predict** that if no women in a country are using contraception, the TFR will be 8.10. Thus:

a is the value of Y when $X = 0$

In other words, a is the value of Y when the regression line crosses the Y axis (this is where $X = 0$). It is sometimes called the Y intercept. In this case the Y intercept is 8.10, so a quick sketch of the regression line would look something like Figure 13.9.

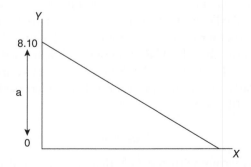

Figure 13.9   A sketch graph showing the regression line and the Y intercept

Be warned, however, that $X = 0$ will not always be a sensible possibility, for example if the variable were the cost of textbooks in a bookshop: none are free! Or again, if the X variable was the age of the woman and the Y variable was the number of children she had: very few (all right then, no) babies aged 0 have children themselves! Always check the context of the data and ask yourself if what you are saying makes sense!

In general, we don't interpret a too much; b is far more interesting:

b is a measure of the 'gradient' or steepness of the line

Suppose you need to estimate the TFR for countries with different percentages of women using contraception. This can be done by putting different percentages into the equation and calculating the resulting TFR. For example, if 10% of women were using contraception, the equation would be:

$$TFR = 8.10 - (0.08 \times 10) = 7.3$$

We would therefore expect the TFR to be 7.3 in a country with 10% contraceptive prevalence.

Table 13.2 shows some more values of the TFR predicted using the equation. Check that you can do these calculations yourself and get the same answers.

Table 13.2 Some values of the total fertility rate estimated from contraceptive prevalence using the regression equation

% women using contraception	Total fertility rate (estimated)
0	8.10
1	8.02
2	7.94
25	6.10
40	4.90
80	1.70
100	0.10

Can you spot what happens every time contraceptive prevalence increases by 1%? The TFR actually falls by 0.08 each time. This is true for all values of contraceptive prevalence, whether it increases from 0% to 1% or 78% to 79%.

Therefore we can say that the TFR increases by b when we increase contraceptive prevalence by 1% (one unit of X). Since b is negative in this case, the TFR actually falls. Similarly, if contraceptive prevalence increased by 10%, the TFR would fall by $10 \times 0.08 = 0.8$.

Warning

Can you see what is wrong with the TFR that we estimated for a country where 100% of women who are married or in a relationship are using contraception? A value of 0.10 for the TFR is unrealistically low (indeed, any childbearing would be the result of contraceptive failure) and would imply that nearly all women are childless! It is also unlikely that 100% of women in relationships would be using contraception in any country, because at any one time, some women would be pregnant or trying to conceive and others would not be fertile.

This problem has arisen because we have estimated from outside the range of our original data. If you look right back to the values we were first given (Table 13.1), you will find that the highest contraceptive prevalence for any of the countries was 84%. We can confidently make predictions about points on the part of the line which comes within the range of the data (12% to 84% of women using contraception). But making predictions outside this range (known as extrapolation) is dangerous! Figure 13.10 shows this.

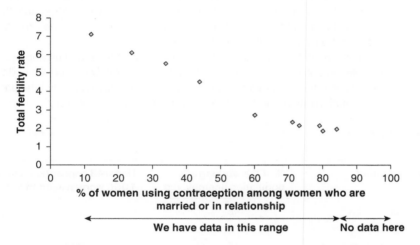

Figure 13.10   Total fertility rate and the percentage of women using contraception among women who are married or in a relationship in 10 countries: the range of the data

Source: World Health Organization, 2012 and United Nations, 2011

Do not make inferences outside the range of your data unless you are really sure that they are sensible. If in doubt, don't!

The regression line: Summary

From the findings above, we can make some general rules which apply to any regression line.

- The equation of a regression line is $Y = a + bX$.
- b is the change in Y for a change of one unit in X.
- If b is positive, Y increases as X increases.
- If b is negative, Y decreases as X increases.
- If $b = 0$, the line will be horizontal and there is no relationship between the two variables.

EXAMPLE: TEENAGE PREGNANCY RATES AND
UNEMPLOYMENT RATES

Teenage conception and early motherhood have been associated with poor educational achievement, social isolation, unemployment and poverty. The Labour government (1997–2010) targeted a reduction in the under-18 conception rate in England. While these targets were discontinued under the coalition government, teenage pregnancy has remained an area of policy interest. Humby (2013) provides data about teenage pregnancy rates (under-18 conceptions per 1,000 women aged 15–17) in different local authorities throughout England and Wales (2008–10). Information is also provided on levels of unemployment among the working-age population in different local authorities. Unemployment has been chosen for analysis as it can be seen as an indicator for deprivation and poverty. It is a domain in the English Indices of Multiple Deprivation and is linked to income. Humby (2013: 35) stated that 'children growing up in areas with high unemployment were likely to experience deprivation and these areas were likely to have high under-18 conception rates'. This will now be explored in relation to a selection of local authorities in London.

Table 13.3 shows under-18 annual conception data (2008–10) and unemployment levels for a selection of local authorities in London. How strongly related are under-18 annual conception rates and unemployment? Can we successfully predict the under-18 annual conception rate of a local authority from its unemployment rate?

Table 13.3 Under-18 conception rate and working-age population unemployment in 10 local authorities in London, 2008–10

Local authority	% of the working age population unemployed	Under-18 conception rate (per 1,000 women aged 15 to 17)
Camden	7.8	31.8
Haringey	10.7	56.0
Kensington and Chelsea	7.0	20.7
Barking and Dagenham	11.1	54.6
Brent	8.9	38.1
Greenwich	9.6	61.3
Kingston upon Thames	5.5	26.2
Richmond upon Thames	4.9	20.2
Sutton	5.7	32.9
Waltham Forest	10.6	50.4

Source: Humby, 2013

Y is the dependent variable, the variable being predicted. In this case the under-18 annual conception rate may be affected by the percentage unemployed, so the under-18 annual conception rate is the Y variable.

The first step as always is to draw a scatter graph, as in Figure 13.11. This shows that, in general, local authorities with high unemployment have high under-18 annual conception rates, while local authorities with low unemployment have low under-18 annual conception rates. According to the literature on this topic, this makes substantive sense.

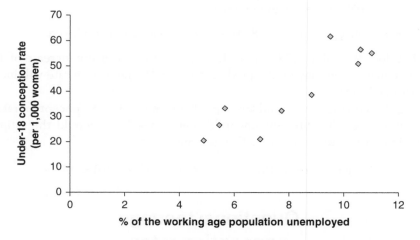

Figure 13.11 Under-18 conception rate and percentage working-age population unemployed in 10 local authorities in London, 2008–10

Source: Humby, 2013

If we were to calculate the regression line for these data, the working would be as follows:

X	Y	$(X_i - \bar{X})$	$(Y_i - \bar{Y})$	$(X_i - \bar{X})^2$	$(X_i - \bar{X})(Y_i - \bar{Y})$
7.8	31.8	−0.38	−7.42	0.14	2.82
10.7	56.0	2.52	−16.78	6.35	42.29
7.0	20.7	−1.18	−18.52	1.39	21.85
11.1	54.6	2.92	15.38	8.53	44.91
8.9	38.1	0.72	−1.12	0.52	−0.81
9.6	61.3	1.42	22.08	2.02	31.35
5.5	26.2	−2.68	−13.02	7.18	34.89
4.9	20.2	−3.28	−19.02	10.76	62.39
5.7	32.9	−2.48	−6.32	6.15	15.67
10.6	50.4	2.42	11.18	5.86	27.06
			Total	48.84	282.42
$\bar{X}$ = 8.18	$\bar{Y}$ = 39.2				

We now have the figures we need to describe our regression line:

$$b = \frac{\Sigma(X - \bar{X})(Y - \bar{Y})}{\Sigma(X - \bar{X})^2} = \frac{282.42}{48.84} = 5.78$$

$$a = \bar{Y} - b\bar{X} = 39.22 - (5.78 \times 8.18) = -8.06$$

So, as $Y = a + bX$, the regression line will be:

Under-18 conception rate = –8.06 + 5.78 percentage unemployed

What does the value 5.78 tell us? It indicates that for every increase of 1% in the percentage unemployed, the predicted under-18 conception rate will increase by 5.78 per 1,000.

Using this equation, we could try to predict some under-18 conception rates for local authorities with different unemployment rates. What would the estimated rate be in a local authority with 10% unemployment?

Under-18 conception rate = –8.06 + (5.78 × 10) = 49.74

Questions: teenage conception rates

Try predicting the under-18 conception rate rankings for local authorities with the following percentages unemployed (answers at the end of the chapter):

(a) 3% unemployed
(b) 7% unemployed
(c) 15% unemployed

Note that your answers will not be the same as the observed rates for local authorities, because we are making a prediction and the observed results are unlikely to be exactly the same as our prediction.

Measuring how well a regression line fits the data

Figure 13.12 shows this regression line plotted on the graph. We can see that most of the points are close to the line. In other words the regression line 'fits' the data well. This means that any predictions we make are likely to be fairly reliable.

In order to measure how well a line fits the data, we must return to r, the correlation coefficient. Remember that when all the points lie exactly on the regression line, the line is a perfect fit and $r = 1$ or $r = -1$. If $r = 0$, then there is no association and the line is useless.

For this purpose, it does not matter whether the correlation is positive or negative; we only want to see whether the line is a good fit. To get rid of the + and – signs we can take the square of r (multiply r by r) and use that as the measure of fit. A value of $r^2 = 1$ indicates a perfect fit, while a value of $r^2 = 0.8$ would be an excellent fit in the social sciences. Even values around $r^2 = 0.5$ are considered, with many social science data, to indicate a good fit. However, lower values would indicate that the line does not fit the data very well.

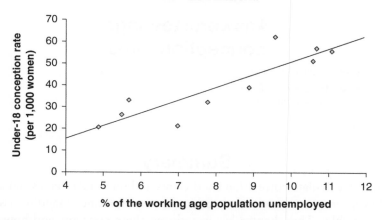

Figure 13.12 Under-18 conception rate and percentage working-age population unemployed in 10 local authorities in London, 2008–10, with a regression line

Source: Humby, 2013

Let's calculate r^2 for under-18 annual conception rates, using the formula for r. Most of the values needed have already been calculated in the working for the regression line, with the exception of

$$\sum(Y - \bar{Y})^2$$

which equals 2101.15 (you could calculate this column yourself for practice). Therefore

$$r = \frac{\sum[(X_i - \bar{X})(Y_i - \bar{Y})]}{\sqrt{\sum(X_i - \bar{X})^2 \sum(Y_i - \bar{Y})^2}} = \frac{282.42}{\sqrt{48.84 \times 2101.15}} = 0.88$$

We square this value to get $r^2 = 0.88^2 = 0.77$.

If the scatter graph does not show a clear association, it is always a good idea to check r^2 first, in order to avoid unnecessary calculation of the regression line.

We have already said that r^2 is a measure of how well the regression line fits the data. Another way of putting this is to say that r^2 *is the proportion of variation in Y which is explained by X*; $1 - r^2$ is therefore the proportion of the variation in Y *not* explained by X.

In this case, 0.77 or 77% of the variation in under-18 annual conception rates is explained by differences in the percentage unemployed. This means that this is an excellent fit. However, there must be other factors involved in determining under-18 annual conception rates which the researcher needs to investigate, such as the accessibility of sexual health clinics.

What about our first example, the total fertility rates and contraceptive prevalence in 10 countries? In Figure 13.11 the data all seem to lie fairly close to the regression line. For these data, we found that $r = -0.99$ and so $r^2 = 0.98$. This indicates that the regression line fits these data extremely well and so we can predict a country's total fertility rate from its contraceptive prevalence with very reliable results. Contraceptive prevalence explains 98% of the difference in total fertility rates between the 10 countries.

Answers: teenage conception rates

(a) Rate $= -8.06 + (5.78 \times 3) = 9.28$
(b) Rate $= -8.06 + (5.78 \times 7) = 32.40$
(c) Rate $= -8.06 + (5.78 \times 15) = 78.64$

Summary

The ability to calculate correlation coefficients and undertake regression analysis is a useful skill for a social researcher to have in terms of understanding associations between variables. This chapter has introduced these concepts and how they can be represented graphically in order to help you understand relationships between variables. Correlation measures the association between two continuous variables, while regression analysis takes things a little further and allows us to predict the values of one variable from the values of another variable. For example, if you had data on your drinking habits on a Saturday night and the number of goals scored in Sunday league matches as a striker the following day, you could carry out a regression analysis in the pub to try to help you decide whether it is worth having that extra pint or it is likely to impact on your goal scoring form the next day! Social statistics really do have their uses!

PRACTICE QUESTIONS

13.1 Students entering higher education are usually vetted by their grades in A-level exams. However, are such grades really good predictors of performance in higher education?

A university decides to look at the question. It allocates points scores for full A-levels, where an A* grade is 12 points, an A grade is 10 points, a B grade 8 points, a C grade 6 points, a D grade 4 points and an E grade 2 points. The number of points for each subject can then be added up to give a points score for each student. Table 13.4 shows the number of points achieved at A level by an anonymous group of students and also their mean mark in their first-semester exams of their first year in higher education.

(a) Which variable will be the dependent variable Y?
(b) Draw a scatter graph of the data. What does it show?
(c) Calculate the correlation coefficient. What does the result mean?
(d) Calculate r^2. How good a predictor of first-semester performance are A-level grades for these students?

13.2 Access to safe drinking water sources is often linked to reduction in illness in developing countries. Table 13.5 shows data on life expectancy in 2009 in 10 countries, along with the percentage of their populations with access to drinking water sources which have been improved to make them safe.

Table 13.4 A-level point scores and mean first semester marks for a group of 18 higher education students

A-level points score[1]	Mean first semester marks (%)
18	54
18	48
16	52
18	65
14	46
12	65
20	52
18	57
20	63
18	53
26	64
16	54
20	43
16	58
28	60
22	64
22	60
22	54

[1]See text for explanation of A-level points scores.

Source: Anonymised student records

Table 13.5 Life expectancy at birth and percentage of population with access to drinking water sources which have been improved to make them safe in 10 countries

Life expectancy at birth, 2009 (years)	% of population using safe drinking water sources, 2010
82	100
75	85
52	51
44	29
72	83
54	44
60	59
49	45
68	82
79	99

Source: World Health Organization, 2012

(Continued)

(Continued)

(a) Draw a scatter graph. What does it tell you about the relationship between access to drinking water sources which have been improved to make them safe and life expectancy?
(b) Calculate the correlation coefficient. What does your answer mean?
(c) Calculate the regression coefficients *a* and *b*.
(d) Write down, in words, the equation for the regression line.
(e) If nobody in a population had access to drinking water sources which have been improved to make them safe, what would we predict life expectancy to be? Is this realistic?
(f) Predict the life expectancy of a country with a percentage of the population with access to drinking water sources which have been improved to make them safe of: (i) 40%, (ii) 52%, (iii) 90%.

FOURTEEN

ANALYSING TABLES WITH CATEGORICAL DATA

Introduction

Much of this book has concentrated on methods for analysing continuous data. This final chapter describes a method for analysing categorical data. You will (hopefully!) remember from Chapter 2 that categorical variables or discrete variables are not measured on a continuous numerical scale. So, categorical variables represent types of data which may be divided into groups such as race, sex, age group or educational level. Chapter 2 also explained that it is possible to further split categorical variables into different levels of measurement: nominal (variables having two or more categories, but not having an intrinsic order or inherent numerical quality), dichotomous (nominal variables which have only two categories or levels) or ordinal (variables having two or more categories like nominal variables but where the categories can also be ordered or ranked). It is probably worth noting that if you are exploring the relationship between two variables at the ordinal level of measurement, there are other options that might be 'better' suited to your data which are beyond the scope of this book. It is also possible to use inferential statistics to analyse categorical data.

One of the most frequently used tests is the **chi-square test**. The chi-square test can be used to find out whether there is a significant relationship between two categorical variables by providing a means of determining whether a set of observed frequencies deviate significantly from a set of expected frequencies. The Greek letter 'chi' is written χ, so in short this is known as the χ^2 test. For those of you have forgotten their maths (or Greek) this means 'the square of chi', pronounced 'kai'. So if you wanted to find out whether income group was associated with home ownership you could perform a chi-square analysis on the frequencies to discover whether there was an association or not. This chapter will introduce you to the chi-square test, when it can be used and how to subsequently analyse and present your findings. It will also show you how residuals can be used to enhance your analysis. By the end of the chapter you should be able to:

- Correctly identify when to use the chi-square test of association
- Calculate the chi-square statistic and assess the significance of the result

- Work out residuals for the cells and use these to analyse associations between variables
- Interpret, report and critically understand the implications of your results

Categorical data and contingency tables

We often want to classify people into different categories. These might be demographic or socio-economic categories such as age, sex or educational qualifications. Other types of survey question also categorise people by their response to a particular question, for example 'yes' or 'no', or perhaps 'strongly agree', 'agree', 'disagree', 'strongly disagree' or 'not sure' about a particular statement. This kind of categorical data consists of **counts** – in other words, the number of people in each category. Tables 14.1 and 14.2 show some data from a hypothetical local survey of 100 people. The people were first asked whether they were aged 'below 40' or '40 or above' and then whether they supported or opposed proposals for a new out-of-town superstore in their area. The data simply show the number of people in each age group and the number of people who supported, opposed or were not sure about the superstore. Tables 14.1 and 14.2 are one-way **contingency tables** or **cross-tabulations** because they contain counts for only one variable.

Table 14.1 Age group of respondents to a local survey

Under 40	40 or above	Total
50	50	100

Source: Hypothetical data

Table 14.2 Opinions of respondents to a local survey

Mainly support superstore	Mainly oppose superstore	Not sure	Total
27	52	21	100

Source: Hypothetical data

Often we may want to see whether there is a relationship between two variables; for example, whether opinions vary by age. To do this, the first step is to produce a two-way contingency table where people are classified by both age and opinion. In Table 14.3, each **cell** contains the number of people of a particular age and opinion. Thus, for example, there were 20 people aged under 40 who mainly support the superstore proposal. Note that the table contains the *row totals* (the totals for each age group: 50 and 50), the *column totals* (the totals for each opinion: 27, 52 and 21) and a *grand total* of 100 people.

You may be able to spot some interesting patterns from Table 14.3. It would appear that those supporting the superstore proposal are mainly under 40, while those opposing the proposal are more likely to be 40 or above. If age had no influence on opinion we might expect the proportion of people supporting and opposing to be similar in both age groups. To investigate this further we could calculate row percentages or column percentages (see Chapter 2 to revise this topic). For example, of the

Table 14.3 Opinions of respondents to a local survey by age group

	Opinion			
Age group	Mainly support superstore	Mainly oppose superstore	Not sure	Total
Under 40	20	17	13	50
40 or above	7	35	8	50
Total	27	52	21	100

Source: Hypothetical data

supporters, 74% ($(20 \div 27) \times 100$) are under 40 and 26% ($(7 \div 27) \times 100$) are 40 or above (this is a column percentage). If age did not affect opinion behaviour we might expect about 50% of the supporters to be under 40 and 50% to be 40 or above, given that we have the same number of respondents in each age group. Therefore it seems that there is some kind of relationship between age and opinion.

However, even if the percentages show an apparent association, we need to test whether this association is significant or not. Does age really influence opinion, or have we obtained these results by chance?

The chi-square test

The chi-square test for independence tests whether two variables are independent of each other. If not, we have evidence that the two variables are associated in some way.

Note that the chi-square test only uses tables of counts. It cannot be used for tables of percentages because the test needs to take into account the total number of people in the sample when determining whether an association is significant. You should not use a chi-square test for tables containing means, proportions or anything else other than counts. The statistic is used when your variables are at the nominal, dichotomous or ordinal levels of measurement – when your data are categorical.

We will start with a worked example using some real data.

EXAMPLE: EDUCATION AND VOTING IN THE 2010 GENERAL ELECTION

The data in Table 14.4 come from an Ipsos MORI survey on voting intention for the 2010 general election carried out between 19 March and 5 May 2010. The sample consists of 4276 individuals who were eligible to vote in 2010 and intended to vote for the Conservatives or Labour. Respondents are classified by their social class and which of the two main parties they intended to vote for in the 2010 election (those people who intended to vote for other parties were excluded from the analysis). Note that the row and column totals are included here, whereas if you had collected the data you would need to work them out yourself first.

Table 14.4 Observed values for social class and voting intention in the 2010 British general election from a sample interviewed between 19 March and 5 May 2010 (only including the main two political parties)

	Political party voted for		
Social class[1]	Conservative	Labour	Total
AB	783	516	1299
C1	739	529	1268
C2	445	349	794
DE	403	532	935
Total	2370	1926	4296

[1]A = High managerial, administrative or professional; B = Intermediate managerial, administrative or professional; C1 = Supervisory, clerical and junior managerial, administrative or professional; C2 Skilled manual workers; D = Semi- and unskilled manual workers; E = State pensioners, casual or lowest grade workers, unemployed with state benefits only.

Source: Ipsos MORI 2010

It would be useful to carry out a test to establish whether there is any relationship between social class and voting behaviour.

Step 1

The first stage in any kind of test is always to write down the null and alternative hypotheses. In this case we might write:

H_0: there is *no* association between social class and party voted for

H_A: there *is* an association between social class and party voted for

In order to decide whether to accept or reject the null hypothesis, we must calculate a test statistic and then compare it to a critical value from the χ^2 tables.

Step 2

Whenever we collect data, they will usually vary from what is expected.

The counts in the original table are known as the **observed values**: they tell us what the survey actually found. For the χ^2 test we need to calculate a table of **expected values**. These are the values which we would expect to get in each cell if there were no association between the two variables. The test then compares the observed and expected values to see whether they are significantly different.

For example, if you roll a dice 600 times it is unlikely that you will roll 100 ones, 100 twos, 100 threes (expected scores), but you are likely to roll something around those scores. Is this just a bit of random variation around these expected values, or are the observed scores due to something other than chance (such as with a loaded dice)?

To calculate the expected value for each cell, we use the formula:

$$\text{Expected value} = \frac{\text{column total} \times \text{row total}}{\text{grand total}}$$

For example, for Conservative voters in social class category AB:

$$\text{Expected value} = \frac{2370 \times 1299}{4296} = 716.63$$

Table 14.5 Expected values for social class and voting intention in the 2010 British general election from a sample interviewed between 19 March and 5 May 2010 (only including the main two political parties)

	Political party voted for		
Social class[1]	Conservative	Labour	Total
AB	716.63	582.37	1299
C1	699.53	568.47	1268
C2	438.03	355.97	794
DE	515.82	419.18	935
Total	2370.00	1926.00	4296

[1] A = High managerial, administrative or professional; B = Intermediate managerial, administrative or professional; C1 = Supervisory, clerical and junior managerial, administrative or professional; C2 Skilled manual workers; D = Semi and unskilled manual workers; E = State pensioners, casual or lowest grade workers, unemployed with state benefits only

Source: Ipsos MORI 2010

Table 14.5 shows the expected values for all six cells. If you have calculated them correctly, the row and column totals should be the same for the expected values as for the observed values: this is a useful check.

Step 3

Now we are ready to calculate the test statistic. The formula for the χ^2 test statistic is as follows:

$$\chi^2 = \Sigma \frac{(O - E)^2}{E}$$

where O represents the observed values and E the expected values.

It is best to do this calculation on a worksheet or a spreadsheet as shown in the following tabulation, where the calculation using O and E is done for each cell and then the results are summed at the end. Once you understand how this works you will be able to miss out some of the columns and just do them on a calculator, but this first example shows the workings in detail.

O	E	(O − E)	(O − E)²	$\frac{(O-E)^2}{E}$
783	716.63	66.37	4404.98	6.15
739	699.53	39.47	1557.88	2.23
445	438.03	6.97	48.58	0.11
403	515.82	−112.82	12728.35	24.68
516	582.37	−66.37	4404.97	7.56
529	568.47	−39.47	1557.88	2.74
349	355.97	−6.97	48.58	0.14
532	419.18	112.82	12728.35	30.36
			Total	73.97

Therefore the χ^2 test statistic in this case is 73.97.

Step 4

But how do we know what a χ^2 test statistic of 73.97 means and whether the value is significant or not? In other words, do our observed scores differ significantly from what we would have expected if the scores were distributed evenly? Answering this question will enable us to decide which hypothesis can be rejected.

Fortunately, we can compare our value against a chart of predetermined critical values. These figures were originally worked out by Karl Pearson (the test is sometimes known as Pearson's chi-square test).

So the next step is to find a critical value from the χ^2 tables (Table A1.3). To do this we must know the significance level we want to use (usually 5%) and the degrees of freedom.

Have a look at the χ^2 table. Along the top is the significance level. To carry out a test at the 5% level we need to look down the column headed χ^2_{df} (0.05). Along the side (in the first column) are the number of degrees of freedom (df). These were first introduced in Chapter 12. Degrees of freedom are important because predetermined critical values largely depend upon the size of the table, so different sizes of table have different critical values.

To find the number of degrees of freedom for a two-way contingency table, we use the following formula:

df = (number of rows − 1) × (number of columns − 1)

In this case the table has four rows and two columns, so the number of degrees of freedom will be $(4 − 1) × (2 − 1) = 3 × 1 = 3$.

Now refer back to the χ^2 table. Look across the row where df = 3 and down the column where the significance level is 5% and you should come to a value of 7.81. This is the critical value of χ^2 for this test.

Step 5

The null hypothesis can be rejected if the test statistic is greater than the critical value, 7.81. The test statistic we calculated in step 3 was 73.97, which is clearly much greater than 7.81. Therefore we can reject the null hypothesis and accept the alternative hypothesis that there is an association between social class and voting intention in the 2010 general election.

The biggest discrepancies in intended voting behaviour between observed and expected scores are in relation to those in the social class categorised as DE (semi- and unskilled manual workers, state pensioners, casual or lowest grade workers, unemployed with state benefits only). In fact if we concentrate on the observed scores this is the only group where the figure for those intending to vote for Labour is higher than for Conservative.

Step 6

Now we know the chi-square critical value and how to interpret it we are in a position to write a statement based on what we have found. For instance:

The study found a significant association between social class and intended voting behaviour ($\chi^2 = 73.97$, df = 3, $p < 0.05$). Examination of the contingency table reveals that the biggest discrepancies in intended voting behaviour between observed and expected scores are in relation to those in the social class categorised as DE (semi- and unskilled manual workers, state pensioners, casual or lowest grade workers, unemployed with state benefits only). This group was more likely to intend to vote for Labour than we would expect if there was no association.

Steps for carrying out a chi-square (χ^2) test

The steps taken in the example are the same for any chi-square test of independence. The following procedure describes how to carry out a chi-square test on any two-way table.

Step 1: Write down the hypotheses

Write down a null hypothesis and alternative hypothesis appropriate to the variables in the table:

H_0: there is no association between the two variables
H_A: there is an association between the two variables

Step 2: Calculate a table of expected values

Remember the formula:

$$\text{Expected value} = \frac{\text{column total - row total}}{\text{grand total}}$$

Always check that your expected values still add up to the same row and column totals as in the original table.

Step 3: Calculate the χ^2 test statistic

The formula is:

$$\chi^2 = \Sigma \frac{(O-E)^2}{E}$$

where O represents the observed values and E the expected values. To calculate this, draw up a worksheet as in the examples, where you calculate $(O-E)^2/E$ for each cell of the table and then sum the results at the end.

Step 4: Find the critical value

Find the critical value from the χ^2 table (Table A1.3). For a test at the 5% level, look down the column headed $\chi^2_{df}(0.05)$. For a test at the 1% level, look down the column headed $\chi^2_{df}(0.01)$. To decide which row to look across, the number of degrees of freedom (df) must be calculated using the formula:

df = (number of rows − 1) × (number of columns − 1)

Finally look down the column with the correct significance level and across the row with the correct number of degrees of freedom and read off the critical value.

With practice, you may become very familiar with some values from the table, for example 3.84 for a 2 × 2 table at the 5% level of significance.

Therefore the rule is:

Test statistic greater than critical value → reject null hypothesis.
Test statistic less than (or equal to) critical value → accept null hypothesis.

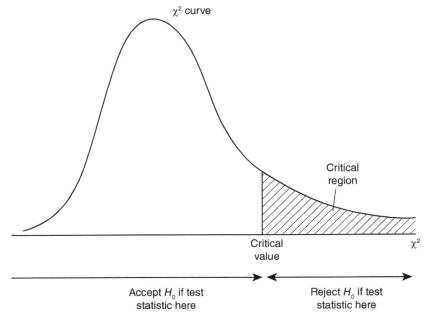

Figure 14.1 **Critical region for** χ^2 **tests**

Some people find it easier to remember the rule; others can more easily memorise the diagram (Figure 14.1).

Step 5: Draw conclusions

Now is the time to make a decision about the hypotheses. The critical region (see Figure 14.1) is the area to the right of the critical value. If the test statistic lies in the critical region, there is evidence to reject the null hypothesis; if not, we must accept it.

Step 6: Write a statement about what you have found

Write a statement based on what you have found from the chi-square statistic. This should clearly identify the critical value, level of significance used and the degrees of freedom, in addition to a description of the differences between observed and expected scores.

EXAMPLE: ARE WOMEN BETTER AT STATISTICS**?**

Table 14.6 shows the results from a Quantitative Methods course taken by students at Southampton University. The students are categorised by gender and by whether they achieved a high mark (55% or better) or a low mark (below 55%). Only students who completed both the coursework and the exam are included here.

The data seem to suggest that there is a greater proportion of men achieving low marks than high marks. Are women significantly better at quantitative methods than men?

Table 14.6   Observed values for Quantitative Methods marks for male and female students undertaking a Quantitative Methods course at Southampton University

| | Sex | | |
Marks	Male	Female	Total
Low[1]	53	53	106
High	36	56	92
Total	89	109	198

[1]'Low' indicates marks of below 55%; 'high' indicates 55% or higher.

Source: Authors' data

Step 1

H_0: there is no association between gender and achievement in Quantitative Methods
H_A: there is an association between gender and achievement in Quantitative Methods

Step 2

Table 14.7 shows the expected values. To calculate the expected value for women with high marks, for example, multiply the row total 92 by the column total 109 and divide your answer by the grand total 198. This gives an expected value of 50.65. Note that we are using two decimal places for accuracy, although whole people are more realistic!

Table 14.7   Expected values for Quantitative Methods marks

| | Sex | | |
Marks	Male	Female	Total
Low[1]	47.65	58.35	106.00
High	41.35	50.65	92.00
Total	89	109	198

Source: Authors' data

Step 3

Next we draw up a worksheet to calculate the test statistic. It does not matter in which order you put the observed values, as long as you put the expected values in the same order!

O	E	(O − E)	(O − E)²	$\dfrac{(O-E)^2}{E}$
53	47.65	5.35	28.62	0.60
36	41.35	−5.35	28.62	0.69

(Continued)

209

(Continued)

O	E	(O − E)	(O − E)²	$\dfrac{(O-E)^2}{E}$
53	58.35	−5.35	28.62	0.49
56	50.65	5.35	28.62	0.57
				Total 2.35

The χ^2 test statistic is 2.35.

Step 4

This table has two rows and two columns so there will be $(2 − 1) \times (2 − 1) = 1$ degree of freedom. From the χ^2 table (Table A1.3), the critical value for a 5% level test with one degree of freedom is 3.84.

Step 5

The test statistic 2.35 is less than the critical value 3.84, so it does not lie in the rejection region. The null hypothesis cannot be rejected, so we must conclude that there is no significant association between gender and achievement in Quantitative Methods.

Step 6

The study did not find a significant association between score in achievement in Quantitative Methods and gender ($\chi^2 = 2.35$, df = 1, $p > 0.05$). Examination of the contingency table reveals limited discrepancies in achievement in Quantitative Methods and gender.

Moving on ...

It should be noted that it is possible to extend the ideas of χ^2 tests to tables with more than two variables. We typically use statistical techniques known as loglinear models to analyse such tables. These techniques are beyond the scope of this book, but they are great fun and we really hope you now feel inspired to move on to even greater heights with your quantitative methods.

Restrictions on the chi-square test

It has already been noted that the χ^2 test can only be used for tables containing counts and the variables must be nominal, dichotomous or ordinal. Each count must also be independent. That is, one person or observation should not contribute more than once to the table (one person = one count). The sample size should be at least 20. In addition, for the results of a χ^2 test to be valid *each expected value should be 5 or above.*

If you have cells with expected values below 5, either you can go and collect more data or you can collapse the categories of your table as shown in the following example. In Table 14.8, the expected values for 15–19-year-olds are both below 5. This is because the sample of 15–19-year-olds is very small, so one solution would be to collect more data from this age group. Alternatively, you could collapse the categories of the table by combining the data for 15–19-year-olds with those from 20–24-year-olds, as in Table 14.9. This gives fewer cells in the table but it means that all the expected values are now well above 5.

Table 14.8 Example of a table with expected values below 5

Agreement with statement		Age group 15–19	20–24	25–29	Total
Agree:	observed	1	15	23	39
	expected	1.56	15.08	22.36	
Disagree/not sure:	observed	2	14	20	36
	expected	1.44	13.92	20.64	
Total		3	29	43	75

Source: Hypothetical data

Table 14.9 Example of a table after collapsing categories

Agreement with statement		Age group 15–24	25–29	Total
Agree:	observed	16	23	39
	expected	16.64	22.36	
Disagree/not sure:	observed	16	20	36
	expected	15.36	20.64	
Total		32	43	75

Source: Hypothetical data

Residuals

The problem with the chi-square test is that it is rather general. It tells us whether there is a difference between the observed findings and what we would have expected, but it doesn't tell us where that difference is or how strong it is. So once we have found a particular association there is more we can do. Having a further look at the contingency table allows us to identify where the major differences are and describe them. In order to do this a good starting point is to return to the differences between the observed and expected scores (the residuals).

A residual is basically a number that expresses the difference between our observed score and the expected score and is part of the chi-square calculation: $O - E$.

While residuals can be used to help to tell us where the substantial differences are in the table, an examination of the residuals can sometimes be problematic, especially if we are working with unstandardised residuals. Used in this raw fashion, we are not necessarily comparing 'like with like', especially where one column or row total has considerably more counts than another. This can lead to under- or over-estimating the effects of a variable. A process of standardising residuals allows us to compare 'like with like' and is reasonably straightforward! The standardised residual for each cell is calculated by dividing the difference of the observed and expected values by the square root of the expected value:

$$r = \frac{O - E}{\sqrt{E}}$$

Let's look back at the first example to show how standardised residuals can be used. In the first example, we concluded that there *was* an association between social class and intended voting behaviour in the 2010 British general election. However, simply concluding that there is an association does not tell us very much. We want to know what kind of association there is: for example, which of the two parties the different social classes are more or less likely to vote for than would be expected if social class did not affect intended voting behaviour.

By calculating standardised residuals we can discover more about the association between social class and intended voting behaviour. A residual r can be calculated for each cell in the table using the formula above. For those with a social class category of AB who voted Conservative, the residual will be:

$$r = \frac{783 - 716.63}{\sqrt{716.63}} = 2.48$$

This and other standardised residuals are shown in Table 14.10.

Table 14.10 Standardised residuals for social class and intended voting example

Social class	Political party voted for	
	Conservative	Labour
AB	2.48	−2.75
C1	1.49	−1.66
C2	0.33	−0.37
DE	−4.97	5.51

Interpreting residuals

What do the residuals in Table 14.10 tell us? First, we can look at whether the residual in each cell is positive or negative. A positive residual indicates that there are more people in the cell than we would expect if there were no association; in other words, the observed value is greater than the expected value. Similarly, a negative residual tells us that there are fewer people than we would expect in that cell.

Table 14.10 shows, for example, that those in social class categories AB (high managerial, administrative or professional, intermediate managerial, administrative or professional) are more likely to vote Conservative than we would expect if there were no association, and less likely to vote Labour. However, those in category DE (semi- and unskilled manual workers, state pensioners, casual or lowest grade workers, unemployed with state benefits only) are more likely to vote Labour than we would expect if there were no association, and less likely to vote Conservative.

Second, we can look at the size of the residual to see whether the difference between observed and expected values is significant. To determine significance at the 5% level, the value of each residual can be compared to 1.96 or −1.96. In practice, 2 and −2 are used for ease. Therefore a residual of greater than 2 or less than −2 is significant, while any values between −2 and 2 are not significant.

In the example, all the residuals for those in categories AB and DE are significant, while the residuals for social class categories C1 and C2 are not significant.

We can conclude that those in social class category AB were significantly more likely to have intended to vote Conservative and significantly less likely to have intended to vote Labour in the 2010 general election than we would expect if social class did not influence voting behaviour. Those in social class categories DE were more likely to intend to vote Labour and less likely to intend to vote Conservative than we might expect if social class did not influence voting behaviour. We could also state that those in social class categories C1 and C2 were more likely to intend to vote Conservative than Labour, but we would need to make it clear that these results are not significant at the 5% level.

Summary

This chapter has introduced you to the use of inferential statistics with categorical data. One of the most common tests is the chi-square test, which can be used to find out whether there is a significant relationship between two categorical variables. If you wanted to find out whether the chance of being in prison in adulthood is associated with a juvenile conviction you could perform a chi-square analysis on the frequencies to determine whether there was an association or not. You should now be able to conduct this analysis if provided with the relevant data! You should also be aware of how to calculate residuals and how these can be used to tell you more about significant associations between variables when using chi-square tests.

PRACTICE QUESTIONS

14.1 A British Social Attitudes Survey asked about attitudes towards mothers' employment when there is a child under school age in 1989 and again in 2012 (Park et al., 2013). Table 14.11 shows the results from this question.

(Continued)

(Continued)

(a) Calculate row and column totals and a grand total for the table. What tentative conclu-
     sions, if any, might you draw from the table?
(b) Carry out a χ^2 test at the 5% level to find out whether there is any association between
     year and attitudes towards mothers' employment. What does this tell you?

Table 14.11   Attitudes to mothers' employment when there
is a child under school age in 1989 and 2012

	Year	
Number agreed a mother should	**1989**	**2012**
work full-time	25	47
work part-time	331	410
stay at home	678	314

Source: Park et al., 2013

14.2 Has interest in politics changed over time? Table 14.12 shows the British Social Attitudes
Survey findings for levels of interest in politics in 1986 and 2012.

Table 14.12   Interest in politics in 1986 and 2012

	Year	
Interest in politics	**1986**	**2012**
Great deal/quite a lot	449	396
Some	480	352
Not much/none at all	604	352

Source: Park et al., 2013

(a) Carry out a χ^2 test at the 5% level to determine whether there is an association between
     the year of the survey and the level of interest in politics.
(b) Would your conclusion change if you carried out the test at the 1% level?
(c) Calculate standardised residuals. What do they tell you about the association between
     the year the survey was carried out and levels of interest in politics?

14.3 Table 14.13 shows some hypothetical data about attitudes towards legalising euthanasia
in Britain by age based on a small-scale student project.

(a) Calculate the expected values. What do you notice?
(b) If you have calculated the expected values correctly, some of them will be below 5. To
     overcome this problem draw up a new contingency table which combines the age
     '55–64' and '65+' categories. Recalculate the expected values to check that they are all
     above 5.

(c) Carry out a χ^2 test at the 5% level on the new table to see whether age has any effect on attitudes towards legalising euthanasia.

Table 14.13 Attitudes towards legalising euthanasia in Britain by age

	Do you think that euthanasia should be legalised in Britain?			
Age	Yes	No	Don't know	Total
18–34	25	10	15	50
35–54	16	15	10	41
55–64	6	10	3	19
65+	2	4	4	10
Total	49	39	32	120

Source: Hypothetical data

FIFTEEN

CONCLUSION

What you (should) now know!

A good grasp of statistics is fundamental to study in any of the social sciences and it is *not* possible to become a social scientist without an understanding of statistics. Even if you think you will never need to calculate any statistics yourself, you will undoubtedly come across statistics in journal articles, books and newspapers and need to know how to interpret them. Now you have worked your way through the book you should feel better prepared to be able to do this. You should hopefully be able to calculate, interpret and present your own statistical findings using a variety of descriptive and inferential statistics too!

This book has included information about how data are collected, measured and transformed, how this affects how they can be presented (including tables, line graphs, histograms, hi-lo plots and scatterplots), how they are distributed and what kind of analysis you can do with them (such as using Z-scores, confidence intervals, two-sided tests and one-sided tests, t tests, correlation coefficients, regression and chi-square tests of significance and residuals). If you have understood these then you will have a good basic grasp of statistics!

So if you noticed that many of your male friends were going to the gym to try to build their muscles while female friends said it was mainly to lose weight and you want to test this statistically, you now know how you could go about doing this. If you wanted to test the association between gender and reasons for attending the gym you should now be aware that they are both categorical variables and that a chi-square test and calculating residuals would be an appropriate technique to use (see Chapter 14 if you have forgotten how to do this!). If you were working on a project about the impact of certain behaviours on health, and smoking in particular, you should now know what you could do! If you are working with continuous data and want to compare two variables, such as number of cigarettes smoked and number of days spent in hospital with lung complaints, you could calculate correlation coefficients. To take this further and predict the impact of smoking 20 a day on the number of days spent in hospital with lung complaints, you would undertake a regression analysis. Hopefully you can see the potential that statistics has to answer important questions across social science disciplines and beyond!

Where do I go from here?

Before you get too excited, it is worth noting that what this book has provided you with is an introduction to using statistics. There is much more that you can do with statistics that you don't yet know about (unless you have done some further reading already!). This book has provided a good foundation (well, we think so!) to learn more about statistics. If you have found this book interesting we would urge you to keep on reading. Many of the statistical tests are no more complicated than we have covered in this book (though some are).

Not choosing the most appropriate statistical test is probably the most common source of error when using statistics (closely followed by not writing results down clearly, failing to back work up and failing to double-check your findings). While this book has introduced you to some criteria that must be fulfilled in order to use particular descriptive and inferential statistics, there are a number of other tests with various requirements that must be met. You can find out more about these techniques in the following texts, some of which are written in a more accessible manner than others. Also note that although some of these state that they are 'introductory' they go into considerable detail. We have confined these to texts based in the social sciences as we believe it is harder to understand statistics if the descriptions constantly refer to areas you are not familiar with or interested in. Anyway, here goes:

Burdess, N. (2010) *Starting Statistics – A Short Clear Guide.* London: Sage.

Dietz, T. and Kalof, L. (2009) *Introduction to Social Statistics.* Oxford: Wiley-Blackwell.

Fielding J. and Gilbert, N. (2006) *Understanding Social Statistics*, 2nd edn. London: Sage.

Frankfort-Nachmias, C. and Leon-Guerrero, A. (2014) *Social Statistics for a Diverse Society*, 7th edn. Thousand Oaks, CA: Sage.

Garner, R. (2010) *The Joy of Stats: A Short Guide to Introductory Statistics for Social Scientists*, 2nd edn. Peterborough: Broadview Press.

Gliner, J., Morgan, G. and Leech, N. (2009) *Research Methods in Applied Settings – An Integrated Approach to Design and Analysis*, 2nd edn. London: Routledge.

Levin, J., Fox, J. and Forde, D. (2013) *Elementary Statistics in Social Research*, 12th edn. London: Allyn & Bacon.

Marsh, C. and Elliott, J. (2008) *Exploring Data: An Introduction to Data Analysis for Social Scientists*, 2nd edn. Cambridge: Blackwell.

Rowntree, D. (2003) *Statistics without Tears: A Primer for Non-Mathematicians*, 2nd edn. London: Allyn & Bacon.

Salkind, N. (2013) *Statistics for People who (Think They) Hate Statistics*, 3rd edn. London: Sage.

Treiman, D. (2009) *Quantitative Data Analysis: Doing Social Research to Test Ideas.* San Francisco: Jossey-Bass.

Walker, J. and Almond, P. (2010) *Interpreting Statistical Findings: A Guide for Health Professionals and Students.* Maidenhead: Open University Press.

Wright, D. and London, K. (2009) *First (and Second) Steps in Statistics*, 2nd edn. London: Sage.

Yang, K. (2010) *Making Sense of Statistical Methods in Social Research*. London: Sage.

Something which we have not mentioned in this book are the computer programs which can help you to undertake statistics, the most common of which used in the social sciences are IBM SPSS, SAS, R and STATA. These allow you to carry out the statistical tests, often at the click of a button, and are particularly useful when dealing with large amounts of data which are accessible in these formats, including national datasets. This book has introduced you to the meaning behind descriptive and inferential statistics and the processes involved in their interpretation. It will be particularly useful when moving on to use statistical computer programs and associated books, which often have a tendency to skip 'much of the theoretical underpinnings of statistics in favour of practically conducting the analysis' (Greasley, 2008: 111). There are a number of excellent books which cover the use of these programmes in the social sciences and which this book will complement, the most detailed of which is Field (2013):

Acton, C., Millar, R. with Fullarton, D. and Maltby, J. (2009) *SPSS for Social Scientists*, 2nd edn. Basingstoke: Palgrave Macmillan.

Babbie, E., Halley, F. and Ziano, J. (2013) *Adventures in Social Research: Data Analysis Using IBM SPSS*, 8th edn. Thousand Oaks, CA: Pine Forge Press.

Bryman, A. and Cramer, D. (2011) *Quantitative Data Analysis with IBM SPSS 17, 18 and 19: A Guide for Social Scientists*. London: Routledge.

Davis, C. (2013) *SPSS Step by Step: Essentials for Social and Political Science*. Bristol: Policy Press.

Faherty, V. (2008) *Compassionate Statistics: Applied Quantitative Analysis for Social Sciences*. Los Angeles: Sage.

Field, A. (2013) *Discovering Statistics Using IBM SPSS for Windows*, 4th edn. London: Sage.

Greasley, P. (2008) *Quantitative Data Analysis Using SPSS: An Introduction for Health and Social Science*. Maidenhead: Open University Press.

Kinnear, P. and Gray, C. (2010) *SPSS 18 Made Simple*. Hove: Psychology Press.

Kulas, J. (2009) *SPSS Essentials*. San Francisco: Jossey-Bass.

Norris, G., Qureshi, F., Howitt, D. and Cramer, D. (2012) *Introduction to Statistics with SPSS for Social Science*. Harlow: Pearson.

Pallant, J. (2013) *SPSS Survival Manual*, 5th edn. Maidenhead: Open University Press.

Final thoughts

It is important to remember that the decisions you make when designing research (for instance, whether you are going to use a categorical or continuous variable) have considerable implications for the analysis you can subsequently undertake. So using statistics doesn't just start when you sit down with your findings and wade through books to work out which statistical tests you could use. It starts with the design, so leave plenty of time for this stage of your work! As Rugg (2007: 122) states, 'statistics are not just about calculations; they're about a way of thinking, and a way of tackling research questions'. This includes how you construct the variables you use and how you phrase your research hypothesis, so it links with the most appropriate statistical test. So while using statistics can save time and money, especially when using secondary data (such as national datasets that have already been collected – remember, though, you don't choose the questions), the use of statistics needs to be carefully thought through and constructed like any form of empirical research, otherwise you may find yourself not effectively answering your research question(s).

Those people who favour different approaches to undertaking social research may accuse research with statistics of suffering from the law of implementation, a view that statistics try to force every area of possible research into the same shape of hole (just as a small child with a hammer tends to believe everything needs to be hammered). If you have read this book (and taken it in) you are unlikely to face the counter-accusation that people who claim this are trying to conceal their inability to use statistics! However, our point here is not to get embroiled in the much covered debates about the pros and cons of statistics and other forms of research such as semi-structured interviews and ethnographies, but to point out that there are usually multiple ways of addressing research questions, which can often be equally useful. In fact, statistics can often be a useful accompaniment to other forms of social research. For instance, Taylor-Gooby (2001) conducted a series of focus groups exploring the idea of risk society and its implications for social welfare and also used data from the first seven waves of the British Household Panel Survey (BHPS) to explore these issues further on a larger scale. He found the focus groups suggested that individuals are aware of the new uncertainties associated with risk societies, while the BHPS showed that while most people viewed the future with some optimism, many have difficulty predicting their future accurately.

Now you have a basic knowledge of statistics you can decide when to use it in your own studies. We hope you have found this book interesting, understand that there is more to statistics than symbols and (hard) calculations and want to use this book as a foundation to build your statistical skills!

APPENDIX 1

STATISTICAL TABLES
THE NORMAL DISTRIBUTION

An area in Table A1.1 is the proportion of the area under the entire curve in Figure A1.1 which lies between the mean (Z = 0) and a positive Z-score. The area between the mean and a negative Z-score will be identical to the area between the mean and a positive Z-score due to symmetry.

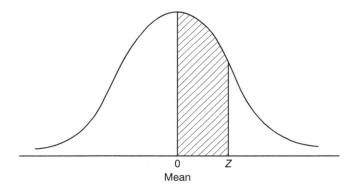

Figure A1.1

Table A1.1 The normal distribution

Z	0.00	0.01	0.02	0.03	0.04	0.05	0.06	0.07	0.08	0.09
0.0	0.0000	0.0040	0.0080	0.0120	0.0160	0.0199	0.0239	0.0279	0.0319	0.0359
0.1	0.0398	0.0438	0.0478	0.0517	0.0557	0.0596	0.0636	0.0675	0.0714	0.0753
0.2	0.0793	0.0832	0.0871	0.0910	0.0948	0.0987	0.1026	0.1064	0.1103	0.1141
0.3	0.1179	0.1217	0.1255	0.1293	0.1331	0.1368	0.1406	0.1443	0.1480	0.1517
0.4	0.1554	0.1591	0.1628	0.1664	0.1700	0.1736	0.1772	0.1808	0.1844	0.1879
0.5	0.1915	0.1950	0.1985	0.2019	0.2054	0.2088	0.2123	0.2157	0.2190	0.2224
0.6	0.2257	0.2291	0.2324	0.2357	0.2389	0.2422	0.2454	0.2486	0.2517	0.2549
0.7	0.2580	0.2611	0.2642	0.2673	0.2704	0.2734	0.2764	0.2794	0.2823	0.2852
0.8	0.2881	0.2910	0.2939	0.2967	0.2995	0.3023	0.3051	0.3078	0.3106	0.3133

Z	0.00	0.01	0.02	0.03	0.04	0.05	0.06	0.07	0.08	0.09
0.9	0.3159	0.3186	0.3212	0.3238	0.3264	0.3289	0.3315	0.3340	0.3365	0.3389
1.0	0.3413	0.3438	0.3461	0.3485	0.3508	0.3531	0.3554	0.3577	0.3599	0.3621
1.1	0.3643	0.3665	0.3686	0.3708	0.3729	0.3749	0.3770	0.3790	0.3810	0.3830
1.2	0.3849	0.3869	0.3888	0.3907	0.3925	0.3944	0.3962	0.3980	0.3997	0.4015
1.3	0.4032	0.4049	0.4066	0.4082	0.4099	0.4115	0.4131	0.4147	0.4162	0.4177
1.4	0.4192	0.4207	0.4222	0.4236	0.4251	0.4265	0.4279	0.4292	0.4306	0.4319
1.5	0.4332	0.4345	0.4357	0.4370	0.4382	0.4394	0.4406	0.4418	0.4429	0.4441
1.6	0.4452	0.4463	0.4474	0.4484	0.4495	0.4505	0.4515	0.4525	0.4535	0.4545
1.7	0.4554	0.4564	0.4573	0.4582	0.4591	0.4599	0.4608	0.4616	0.4625	0.4633
1.8	0.4641	0.4649	0.4656	0.4664	0.4671	0.4678	0.4686	0.4693	0.4699	0.4706
1.9	0.4713	0.4719	0.4726	0.4732	0.4738	0.4744	0.4750	0.4756	0.4761	0.4767
2.0	0.4772	0.4778	0.4783	0.4788	0.4793	0.4798	0.4803	0.4808	0.4812	0.4817
2.1	0.4821	0.4826	0.4830	0.4834	0.4838	0.4842	0.4846	0.4850	0.4854	0.4857
2.2	0.4861	0.4864	0.4868	0.4871	0.4875	0.4878	0.4881	0.4884	0.4887	0.4890
2.3	0.4893	0.4896	0.4898	0.4901	0.4904	0.4906	0.4909	0.4911	0.4913	0.4916
2.4	0.4918	0.4920	0.4922	0.4925	0.4927	0.4929	0.4931	0.4932	0.4934	0.4936
2.5	0.4938	0.4940	0.4941	0.4943	0.4945	0.4946	0.4948	0.4949	0.4951	0.4952
2.6	0.4953	0.4955	0.4956	0.4957	0.4959	0.4960	0.4961	0.4962	0.4963	0.4964
2.7	0.4965	0.4966	0.4967	0.4968	0.4969	0.4970	0.4971	0.4972	0.4973	0.4974
2.8	0.4974	0.4975	0.4976	0.4977	0.4977	0.4978	0.4979	0.4979	0.4980	0.4981
2.9	0.4981	0.4982	0.4982	0.4983	0.4984	0.4984	0.4985	0.4985	0.4986	0.4986
3.0	0.4987	0.4987	0.4987	0.4988	0.4988	0.4989	0.4989	0.4989	0.4990	0.4990

$Z = 3.5$: proportion = 0.4998

$Z = 4.0$: proportion = 0.49997

Source: adapted from Fisher and Yates, 1974; *Statistical Tables for Biological, Agricultural and Medical Research*. Reprinted with the permission of Pearson Education Limited

The *t* distribution

Table A1.2 gives critical values for the *t* distribution. The left-hand column gives the number of degrees of freedom, and the remaining columns give the values for various significance levels, i.e. the proportion of the area in one tail of the distribution in Figure A1.2.

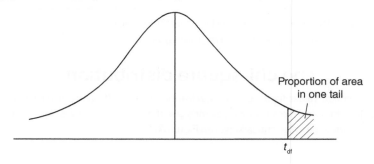

Proportion of area in one tail

t_{df}

Figure A1.2

Table A1.2 The *t* distribution

df	$t_{df}(0.1)$	$t_{df}(0.05)$	$t_{df}(0.025)$	$t_{df}(0.01)$	$t_{df}(0.005)$
1	3.0777	6.3138	12.7062	31.8205	63.6567
2	1.8856	2.9200	4.3027	6.9646	9.9248
3	1.6377	2.3534	3.1824	4.5407	5.8409
4	1.5332	2.1318	2.7764	3.7469	4.6041
5	1.4759	2.0150	2.5706	3.3649	4.0321
6	1.4398	1.9432	2.4469	3.1427	3.7074
7	1.4149	1.8946	2.3646	2.9980	3.4995
8	1.3968	1.8595	2.3060	2.8965	3.3554
9	1.3830	1.8331	2.2622	2.8214	3.2498
10	1.3722	1.8125	2.2281	2.7638	3.1693
11	1.3634	1.7959	2.2010	2.7181	3.1058
12	1.3562	1.7823	2.1788	2.6810	3.0545
13	1.3502	1.7709	2.1604	2.6503	3.0123
14	1.3450	1.7613	2.1448	2.6245	2.9768
15	1.3406	1.7531	2.1314	2.6025	2.9467
16	1.3368	1.7459	2.1199	2.5835	2.9208
17	1.3334	1.7396	2.1098	2.5669	2.8982
18	1.3304	1.7341	2.1009	2.5524	2.8784
19	1.3277	1.7291	2.0930	2.5395	2.8609
20	1.3253	1.7247	2.0860	2.5280	2.8453
21	1.3232	1.7207	2.0796	2.5176	2.8314
22	1.3212	1.7171	2.0739	2.5083	2.8188
23	1.3195	1.7139	2.0687	2.4999	2.8073
24	1.3178	1.7109	2.0639	2.4922	2.7969
25	1.3163	1.7081	2.0595	2.4851	2.7874
26	1.3150	1.7056	2.0555	2.4786	2.7787
27	1.3137	1.7033	2.0518	2.4727	2.7707
28	1.3125	1.7011	2.0484	2.4671	2.7633
29	1.3114	1.6991	2.0452	2.4620	2.7564
30	1.3104	1.6973	2.0423	2.4573	2.7500
∞[1]	1.2816	1.6449	1.9600	2.3263	2.5758

[1] df = ∞ (last row of table) gives critical values of the normal (*Z*) distribution.

(.2, .1, .05, .02 and .01 from Table III: Distribution of *t* Probability p. 46).

The chi-square distribution

Table A1.3 gives critical values for the chi-square distribution. The left-hand column gives the number of degrees of freedom, and the remaining columns give the values for various significance levels, i.e. the proportion of the area in the critical region in Figure A1.3.

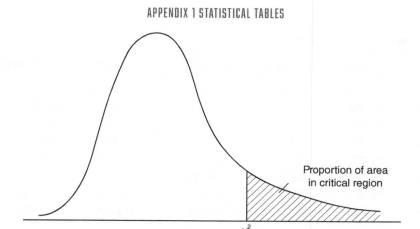

Figure A1.3

Table A1.3 The chi-square distribution

df	$\chi^2_{df}(0.1)$	$\chi^2_{df}(0.05)$	$\chi^2_{df}(0.025)$	$\chi^2_{df}(0.01)$	$\chi^2_{df}(0.005)$
1	2.7055	3.8415	5.0239	6.6349	7.8794
2	4.6052	5.9915	7.3778	9.2103	10.5966
3	6.2514	7.8147	9.3484	11.3449	12.8382
4	7.7794	9.4877	11.1433	13.2767	14.8603
5	9.2364	11.0705	12.8325	15.0863	16.7496
6	10.6446	12.5916	14.4494	16.8119	18.5476
7	12.0170	14.0671	16.0128	18.4753	20.2777
8	13.3616	15.5073	17.5345	20.0902	21.9550
9	14.6837	16.9190	19.0228	21.6660	23.5894
10	15.9872	18.3070	20.4832	23.2093	25.1882
11	17.2750	19.6751	21.9200	24.7250	26.7568
12	18.5493	21.0261	23.3367	26.2170	28.2995
13	19.8119	22.3620	24.7356	27.6882	29.8195
14	21.0641	23.6848	26.1189	29.1412	31.3193
15	22.3071	24.9958	27.4884	30.5779	32.8013
16	23.5418	26.2962	28.8454	31.9999	34.2672
17	24.7690	27.5871	30.1910	33.4087	35.7185
18	25.9894	28.8693	31.5264	34.8053	37.1565
19	27.2036	30.1435	32.8523	36.1909	38.5823
20	28.4120	31.4104	34.1696	37.5662	39.9968
21	29.6151	32.6706	35.4789	38.9322	41.4011
22	30.8133	33.9244	36.7807	40.2894	42.7957

(Continued)

Table A1.3 (Continued)

df	$\chi^2_{df}(0.1)$	$\chi^2_{df}(0.05)$	$\chi^2_{df}(0.025)$	$\chi^2_{df}(0.01)$	$\chi^2_{df}(0.005)$
23	32.0069	35.1725	38.0756	41.6384	44.1813
24	33.1962	36.4150	39.3641	42.9798	45.5585
25	34.3816	37.6525	40.6465	44.3141	46.9279
26	35.5632	38.8851	41.9232	45.6417	48.2899
27	36.7412	40.1133	43.1945	46.9629	49.6449
28	37.9159	41.3371	44.4608	48.2782	50.9934
29	39.0875	42.5570	45.7223	49.5879	52.3356
30	40.2560	43.7730	46.9792	50.8922	53.6720

Source: adapted from Fisher and Yates, 1974; *Statistical Tables for Biological, Agricultural and Medical Research.* Reprinted with the permission of Pearson Education Limited (columns .10, .05 and .01 up to row 30 from Table IV: Distribution of χ^2 Probability, p. 47).

APPENDIX 2
ANSWERS TO PRACTICE QUESTIONS

Chapter 2

2.1   Your table should look something like Table A2.1. The biggest gender differences shown in the table are that a greater proportion of males than females are single, and a much larger proportion of females than males are widowed. The latter probably reflects the fact that women tend to live longer than men. A slightly higher proportion of females than males are divorced, which may be due to the higher remarriage chances for men than women.

Table A2.1   Table to show whether there are any differences in marital status between males and females (all ages) in England and Wales (column %), 2010

	Sex	
Marital status	**Male**	**Female**
Single[1]	51.85	44.84
Married[2]	38.58	37.80
Widowed[3]	2.65	8.64
Divorced[4]	6.92	8.72
Total %	100.00	100.00
Total number (thousands)	27,228.60	28,012.00

[1]People who have never been legally married
[2]People who are currently legally married (including those who are separated)
[3]People who are legally married until the death of their partner, and have subsequently neither remarried nor divorced
[4]People who are legally married but have been legally divorced, or had their marriage annulled, and have not since remarried

Source: Office for National Statistics, 2011

2.2   (a)   Your table should look something like Table A2.2.
       (b)   (i) 28%.
              (ii) 9%.

Table A2.2   Voting behaviour by newspaper readership in the 2010 general election (%)

Newspaper read	Party voted for				Total % (n)
	Conservative	Labour	Lib Dem	Other party	
Mirror	16	59	17	8	100 (534)
Daily Telegraph	70	7	19	4	100 (508)
Guardian	9	46	37	8	100 (530)
Sun	43	28	18	11	100 (988)

Source: Worcester and Herve, 2010

(iii) Those reading the *Daily Telegraph* and the *Sun* were considerably more likely to vote for the Conservative Party than the Labour Party at the 2010 general election, while those reading the *Mirror* and the *Guardian* were most likely to vote for Labour and least likely to vote for the Conservative Party at the 2010 general election. Readers of the *Guardian* were the most likely out of the four newspaper readership groups to vote for the Liberal Democrats at the 2010 general election, while readers of the *Sun* were most likely out of the four newspaper readership groups to vote for another party at the 2010 general election.

Chapter 3

3.1   (a)   An example of an abridged frequency table is shown in Table A3.1. Intervals of a width of 10 are probably most appropriate here.

Table A3.1   Percentage of 1-year-olds fully immunised against measles in African countries in 2010 among the lowest wealth quintile

% fully immunised against measles	Frequency (number of countries)
0–9	3
10–19	1
20–29	3
30–39	3
40–49	1
50–59	6
60–69	5
70–79	6
80–89	5
90–99	3

Source: World Health Organization, 2012

3.1   (b)   An example of a histogram is shown in Figure A3.1. The distribution of the data shows a high proportion of the countries with 50% or more of 1-year-olds fully immunised against measles in 2010 among the lowest wealth quintile. However, there are still a few countries with a lower percentage than this.

3.2   Figure A3.2 is a back-to-back stem and leaf plot showing the percentage of men and women who smoke in 22 industrialised countries in 2009. It is clear that there were a higher percentage

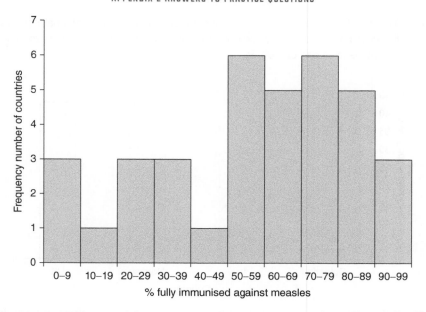

Figure A3.1 A histogram to show the percentage of 1-year-olds fully immunised against measles in African countries in 2010 among the lowest wealth quintile

Source: World Health Organization, 2012

of men smoking than women in the 22 industrialised countries. In five countries 40% or more of men smoked, compared with just one country for women. In five countries less than 20% of women smoked, while there were no countries where the percentage of men smoking was under 20%. For women, the greatest number of countries fall in the 20–24% and 25–29% intervals, while for men the greatest number fall into the 30–34% category.

Males		Females
	1	2
	1*	7 9 9 6
0 4	2	1 2 4 2 3 1 4
5 7 8 7	2*	7 8 5 6 7 8 5
2 3 0 1 0 1 3 1	3	1 3
6 6 6	3*	
3 3 2	4	1
6	4*	
	5	
	5*	
3	6	
	6*	

Figure A3.2 A back-to-back stem and leaf plot showing the percentage of men and women who smoke in 22 industrialised countries, 2009

227

3.3 (a)   The data from Table 3.17 would be best presented in the form of a bar chart (see Figure A3.3). A pie chart would not be suitable because the different percentages do not make up part of a whole.

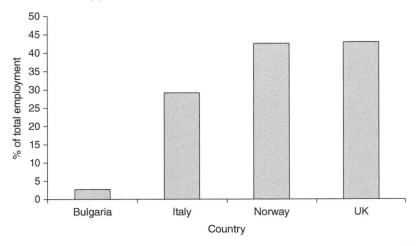

Figure A3.3   A bar chart to show the percentage of women's employment which is part-time in four EU countries, 2011

Source: Eurostat, 2011

3.3 (b)   The data from Table 3.18 would be best presented in the form of a line graph (see Figure A3.4). This enables two lines, one for the UK and one for Spain, to be plotted on the same graph.

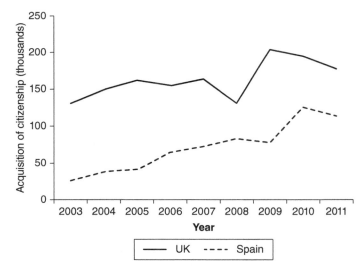

Figure A3.4   A line graph to show the acquisition of citizenship in the UK and Spain, 2003–11 (thousands)

3.3 (c)   The data from Table 3.19 would be best presented in the form of a multiple bar chart, with one colour for the percentage of school leavers by region to higher education and another for the Russell Group universities (see Figure A3.5).

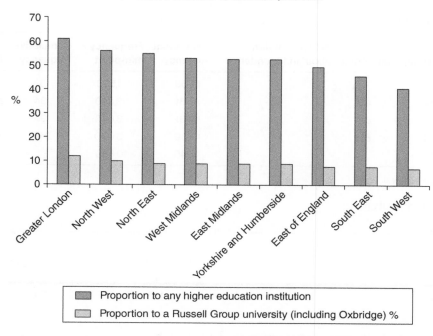

Figure A3.5 **A multiple bar chart to show the known first destination of school leavers by region to higher education and whether it is to a Russell Group university, 2009–10**

Source: Department for Education, 2012

Chapter 4

4.1 (a) The mean is 55,747 ÷ 10 = 5,574.7. This is the mean Scottish Third Division attendance in 2012–13.
The median is the 5.5th observation = (917 + 937) ÷ 2 = 927. This is the median Scottish Third Division attendance in 2012–13.
   (b) Rangers football club is a clear outlier.
   (c) If Rangers is removed from the data:
Mean = 10,003 ÷ 9 = 1,111.4 (1,111 people)
Median = 5th observation = 917

The mean changes more than the median when Rangers is removed because the mean is affected more by extreme values.

4.2 Here the methods for grouped data must be used. Columns were added to Table 4.11 to assist with this process. See Table A4.1.

Mean = Sum of (mid-point × frequency) ÷ Sum of frequencies = 44.41 years (44 years)

Median = (245 + 1) ÷ 2 = 123rd observation.
This lies in the 35–44 age group.

Table A4.1   Confidence in independence by age, Scotland, 2011

Age group	Mid-point	% feel confident about independence	Respondent frequency	Frequency × mid-point	Cumulative frequency	Total
25–34	30	37	53	1,590	53	143
35–44	40	38	81	3,240	134	212
45–54	50	27	61	3,050	195	227
55–64	60	26	50	3,000	245	193
Total			245	10,880		

Source: Park *et al.*, 2012

From interpolation:

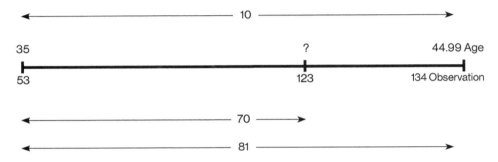

Median = 35 + (70/82 × 10) = 43.54 (44 years)

4.3 (a) £42,754.40 – £12,469.60 = £30,284.80
     (b) £12,469.60
     (c) £7,753.20 – £1,336.40 = £6,416.80
     (d) £1,050.40 – £795.60 = £254.80

4.4 The cumulative frequencies must be calculated first.

   (a) Median = 52nd observation = 2 hrs of soap watching per week
   (b) LQ = 26th observation = 1 hr of soap watching per week
   (c) UQ = 78th observation = 4 hrs of soap watching per week

Chapter 5

5.1 (a) Assessment 1: mean = 556 ÷ 10 = 55.6
          Assessment 2: mean = 543 ÷ 10 = 54.3

The students did not improve between the two assessments: the mean mark fell slightly.

   (b) Assessment 1 – Workings:

$$SD = \sqrt{\frac{208.40}{10-1}} = 4.81$$

X_i	$\bar{X}$	$X_i - \bar{X}$	$\left(X_i - \bar{X}\right)^2$
53	55.6	-2.6	6.76
61	55.6	5.4	29.16
54	55.6	-1.6	2.56
59	55.6	3.4	11.56
59	55.6	3.4	11.56
48	55.6	-7.6	57.76
53	55.6	-2.6	6.76
49	55.6	-6.6	43.56
60	55.6	4.4	19.36
60	55.6	4.4	19.36
Total			208.40

Standard deviation = 4.81

Assessment 2 – Workings:

X_i	$\bar{X}$	$X_i - \bar{X}$	$\left(X_i - \bar{X}\right)^2$
55	54.3	0.7	0.49
77	54.3	22.7	515.29
57	54.3	2.7	7.29
70	54.3	15.7	246.49
61	54.3	6.7	44.89
50	54.3	-4.3	18.49
41	54.3	-13.3	176.89
60	54.3	5.7	32.49
17	54.3	-37.3	1,391.29
55	54.3	0.7	0.49
Total			2,434.10

$$SD = \sqrt{\frac{2434.10}{10-1}} = 16.45$$

Standard deviation = 16.45

The students' performance was much more variable in the second assessment because the standard deviation is much larger.

5.2 (a) The data must first be ordered:

0 0 0 0 0 0 0 0 0 0 1 1 3 4 5 6 12 14 22 28

$$\text{Median} = \frac{n+1}{2} \text{th observation}$$

$$= 21 \div 2 = 10.5\text{th observation}$$

10th observation = 0; 11th observation = 1
Median number of days marijuana is smoked = 0.5

$$\text{Upper quartile} = \frac{3 \times (n+1)}{4} \text{ th observation}$$

$$= 15.75\text{th observation}$$

15th observation = 5; 16th observation = 6
Upper quartile = 5.75 days

$$\text{Lower quartile} = \frac{1 \times (n+1)}{4} \text{ th observation}$$

$$= 5.25\text{th observation}$$

5th observation = 0; 6th observation = 0
Lower quartile = 0 days
Inter-quartile range = upper quartile – lower quartile = 5.75 – 0 = 5.75 days

(b) These data are very skewed. They are positively skewed because most of the students in the sample did not smoke marijuana at all or for very few days in the last 28
days, while four students, two in particular, had smoked marijuana on a large number
of days in the last 28 days. When data are skewed it is better to use the median and
inter-quartile range as measures of the average and spread than to use the mean and
standard deviation.

5.3   A box plot is shown in Figure A5.1.

Partial workings for calculating the boxplot are as follows:
Ordered data:

| 10 | 15 | 20 | 21 | 26 | 29 | 30 | 31 | 32 | 34 | 36 | 37 | 39 | 44 | 48 |
| 50 | 53 | 54 | 55 | 58 | 60 | 61 | 64 | 72 | 122 | 130 | 141 | 185 | 208 | 208 |

Median = (30 + 1) ÷ 2 = 15.5th observation = 49 minutes
Upper quartile = 23.25th observation = 66 minutes
Lower quartile = 7.75th observation = 30.75 minutes
Inter-quartile range = 66 – 30.75 = 35.25 minutes
Lower fence   = lower quartile – (1.5 × inter-quartile range)
                   = 30.75 – (1.5 × 35.25)
                   = 30.75 – 52.875 = –22.125
First observation above lower fence = 10 minutes
Upper fence   = upper quartile + (1.5 × inter-quartile range)
                   = 66 + (1.5 × 35.25)
                   = 66 + 52.875 = 118.875 minutes
First observation below upper fence = 113 minutes
Outliers are 122, 130, 141, 185 and 208 minutes

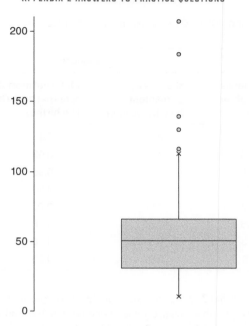

Figure A5.1 Waiting times before being treated in an accident and emergency unit (hypothetical data) (minutes)

Source: Hypothetical data

Comment: The distribution is positively skewed. Half of the people waiting to be treated in an accident and emergency unit wait for less than 49 minutes and three-quarters for less than 66 minutes. The hospital is likely to be particularly concerned about the five outliers taking between 122 and 208 minutes before they are seen.

Chapter 6

6.1 (a) Mean price = 3400 ÷ 15 = 227p

$$\text{Standard deviation} = \sqrt{154495 \div 14} = \sqrt{11035} = 105p$$

(b) Mean price = (227 + 20) = 247p
Standard deviation = 105p (no change)

(c) Mean price = 247 × 120% = 296p
Standard deviation = 105 × 120% = 126p

6.2 Answers are given to one decimal place. The mean and standard deviation for each variable are as follows:

Women as % of labour force: mean = 36.1; SD = 8.8
% seats in parliament held by women: mean = 17.0; SD = 5.0

Maternal mortality rate (per 100,000 live births): mean = 141.6; SD = 98.1

Country	Women as % of labour force	% seats in parliament held by women	Maternal mortality rate (per 100, 000 live births)	Mean Z-score
		Z-scores		
Philippines	0.30	1.01	0.43	0.58
India	–1.23	–1.23	–0.59	–1.01
Thailand	1.09	–0.73	0.95	0.43
Malaysia	–0.03	–1.41	1.14	–0.10
Indonesia	0.23	0.19	–0.79	–0.12
China	0.96	0.85	1.06	0.95
Bangladesh	0.43	0.31	–1.00	–0.08
Pakistan	–1.75	1.03	–1.20	–0.64

The above working shows the Z-scores for each variable. The signs for the maternal mortality variable have been reversed so that positive values become negative and negative values become positive. The final column shows the mean Z-score for each country.

Comment: A high value indicates high female empowerment, while a low index value indicates low female empowerment. From the three variables used it appears that Chinese women are the most empowered, followed by women from the Philippines and Thailand. Women are least empowered in India and Pakistan.

Chapter 7

7.1 (a) 68.2% of the observations lie within one standard deviation either side of the mean; in other words, between 85 and 115.

(b) See Figure A7.1.

$$Z = \frac{80 - 100}{15} = -1.33$$

Area between Z = –1.33 and mean is 0.4082.
Area to left of Z = –1.33 is 0.5 – 0.4082 = 0.0918.
The proportion of the population with an IQ below 80 is 0.0918.

(c) See Figure A7.2.

$$Z = \frac{110 - 100}{15} = 0.67 \, (\text{to two decimal places})$$

Area between mean and Z = 0.67 is 0.2486.

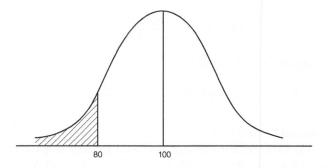

Figure A7.1 Proportion of the population with an IQ below 80

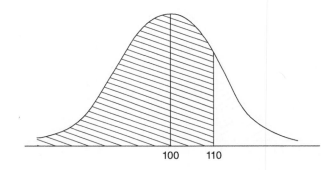

Figure A7.2 Proportion of the population with an IQ below 110

Add the 0.5 to the left of the mean: 0.5 + 0.2486 = 0.7486.
The proportion of the population with an IQ below 110 is 0.7486.

(d) See Figure A7.3.

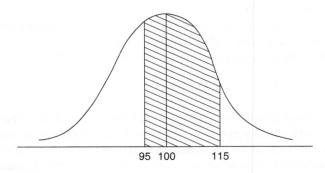

Figure A7.3 Proportion of the population with an IQ between 95 and 115

235

$$\text{For the } 95: Z = \frac{95 - 100}{15} = -0.33$$

Area between mean and Z = –0.33 is 0.1293.

$$\text{For the } 115: Z = \frac{115 - 100}{15} = 1$$

Area between mean and Z = 1 is 0.3413.

The areas are on opposite sides of the mean, so they can be added together: 0.1293 + 0.3413 = 0.4706.
The proportion of the population with an IQ between 95 and 115 is 0.4706.

(e) See Figure A7.4.

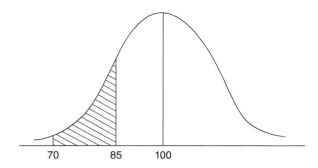

Figure A7.4   **Proportion of the population with an IQ between 70 and 85**

$$\text{For the } 70: Z = \frac{70 - 100}{15} = -2$$

Area between mean and Z = –2 is 0.4772.

$$\text{For the } 85: Z = \frac{85 - 100}{15} = -1$$

Area between mean and Z = –1 is 0.3413.

The areas are on the same side of the mean (both below the mean) and so the smaller area should be subtracted from the larger area: 0.4772 – 0.3413 = 0.1359.
The proportion of the population with an IQ between 70 and 85 is 0.1359.

(f) See Figure A7.5.

We cannot look up the Z-score for the top 10% directly, but we know that 50% of the observations lie above the mean and we are able to look up the Z-score for the 40% just above the mean. The area 0.4 is looked up in the middle of the tables; the nearest is 0.3997, which gives a Z-score of 1.28. This can then be substituted into the equation for Z, where we know the Z-score, the mean and the standard deviation, and the equation is then rearranged to find the answer:

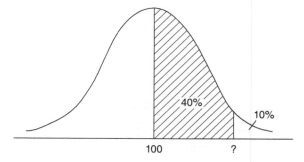

Figure A7.5   Finding the IQ of the top 10% of the population

$$1.28 = \frac{?-100}{15}$$

? = (1.28 ×15) + 100 = 119.2.

The top 10% of the population will have an IQ greater than 119.2.

7.2   (a)   See Figure A7.6.

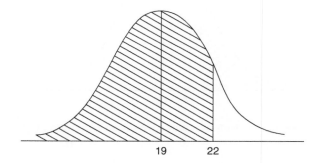

Figure A7.6   Proportion of students taking less than 22 minutes to reach the campus

$$Z = \frac{22-19}{3} = 1$$

Area between mean and Z = 1 is 0.3413.

Add 0.5 for the area below the mean: 0.5 + 0.3413 = 0.8413.

The proportion of students taking less than 22 minutes is 0.8413.

(b) See Figure A7.7.

$$Z = \frac{15-19}{3} = -1.33$$

Area between mean and Z = –1.33 is 0.4082.

Add 0.5 for the area above the mean: 0.5 + 0.4082 = 0.9082.

The proportion of students who take more than 15 minutes is 0.9082.

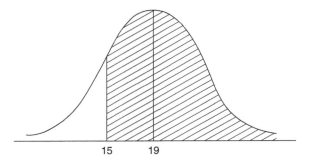

Figure A7.7 Proportion of students taking more than 15 minutes to reach the campus

(c) See Figure A7.8.

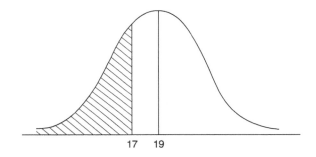

Figure A7.8 Proportion of students taking less than 17 minutes to reach the campus

$$Z = \frac{17 - 19}{3} = -0.67$$

Area between mean and Z = –0.67 is 0.2486.
Area to the left of Z = –0.67 is: 0.5 – 0.2486 = 0.2514.
The proportion of students who take less than 17 minutes is 0.2514.

(d) Here we need to find out the proportion of students who take more than 20 minutes as these students are the ones who will be late for the lecture. See Figure A7.9.

$$Z = \frac{20 - 19}{3} = 0.33$$

Area between mean and Z = 0.33 is 0.1293.
Area to the right of Z = 0.33 is: 0.5 – 0.1293 = 0.3707.
The proportion of students who would be late is 0.3707.

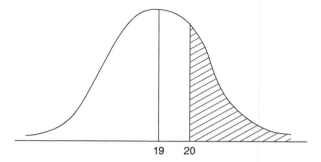

Figure A7.9 Proportion of late arriving students allowing 20 minutes to reach the campus

(e) See Figure A7.10.

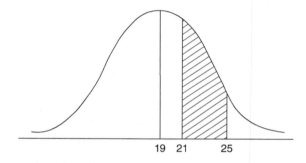

Figure A7.10 Proportion of students taking between 21 and 25 minutes to reach the campus

$$\text{For the } 21: Z = \frac{21-19}{3} = 0.67$$

Area between mean and Z = 0.67 is 0.2486.

$$\text{For the } 25: Z = \frac{25-19}{3} = 2$$

Area between mean and Z = 2 is 0.4772.
The two areas are on the same side of the mean (both above the mean), so the smaller area should be subtracted from the larger area: 0.4772 – 0.2486 = 0.2286.
The proportion of students who take between 21 and 25 minutes to reach the campus is 0.2286.

7.3 (a) See Figure A7.11.

$$Z = \frac{35-58}{10} = -2.3$$

Area between mean and Z = –2.3 is 0.4893.
Area to the left of Z = –2.3 is: 0.5 – 0.4893 = 0.0107.
The proportion of students who fail the exam is 0.0107.

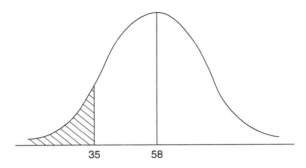

Figure A7.11 Proportion of students failing an exam

(b)   See Figure A7.12.

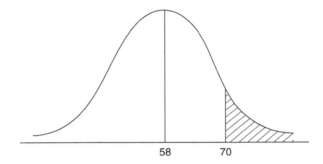

Figure A7.12 Proportion of students awarded an A grade in an exam

$$Z = \frac{70 - 58}{10} = 1.2$$

Area between mean and Z = 1.2 is 0.3849.
Area to the right of Z = 1.2 is: 0.5 – 0.3849 = 0.1151.
The proportion of students who get an A grade is 0.1151.

(c)   We cannot directly find the Z-score corresponding to the top 40%, but we know that 50% of students will get a mark above the mean and we can find the Z-score for the 10% just above the mean (see Figure A7.13). For an area of 0.1, the nearest Z-score is 0.25 (for the area 0.0987). This can now be put into the Z-score equation and rearranged:

$$0.25 = \frac{? - 58}{10}$$

$$? = (0.25 \times 10) + 58 = 60.5$$

Therefore 40% of the students obtained a mark higher than 60.5%.

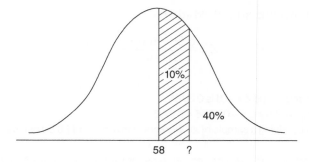

Figure A7.13   Lowest mark of the best 40% of students

Chapter 8

8.1   (a)   See Figure A8.1.

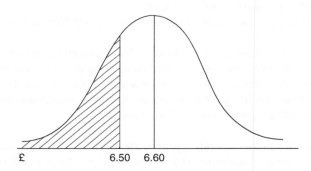

Figure A8.1   **Probability of a sample of 1, 20 or 50 workers earning £6.50 or less per hour**

$$Z = \frac{6.50 - 6.60}{0.40} = -0.25$$

Area between mean and Z = –0.25 is 0.0987.
Area to left of Z = –0.25 is: 0.5 – 0.987 = 0.4013.
The probability of an individual worker earning £6.50 or less per hour is 0.4013.

(b)   The diagram will be the same as for part (a), i.e. Figure A8.1.

Samples of 20 workers will be normally distributed with a mean of £6.60 and a standard error of:

$$SE = \frac{SD}{\sqrt{n}} = \frac{0.40}{\sqrt{20}} = 0.09$$

241

Now Z can be calculated using the standard error:

$$Z = \frac{6.50 - 6.60}{0.09} = -1.11$$

Area between the mean and Z = –1.11 is 0.3665.
Area to the left of Z = –1.11 is: 0.5 – 0.3665 = 0.1335.
The probability of obtaining a sample of 20 workers with a mean of £6.50 or less is 0.1335.

(c)   The diagram will again be the same as for part (a), i.e. Figure A8.1. Samples of 50 workers will be normally distributed with a mean of £6.60 and a standard error of:

$$SE = \frac{SD}{\sqrt{n}} = \frac{0.40}{\sqrt{50}} = 0.06$$

$$Z = \frac{6.50 - 6.60}{0.06} = -1.67$$

Area between the mean and Z = –1.67 is 0.4525.
Area to the left of Z = –1.67 is: 0.5 – 0.4525 = 0.0475.
The probability of obtaining a sample of 50 workers with a mean of £6.50 or less is 0.0475.

(d)   The chance of individuals in a population having a particularly high or low score is usually fairly high. However, the chance of a sample of people having such an extreme mean score is always lower because a mean will never be as extreme as the observations. Sample means vary much less than individual scores. This is why the probability in part (b) is lower than in part (a).

The larger the sample size, the more representative the sample should be of the population and the less likely you are to get an extreme mean by chance. This is why the probability in part (c) is lower than the probability in part (b): the sample size is larger.

8.2 See Figure A8.2.

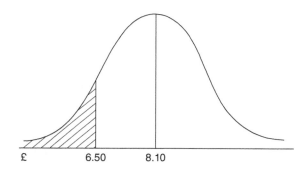

Figure A8.2   Probability of a sample of 30 workers earning £6.50 or less per hour

Samples of size 30 will be normally distributed with a mean of £8.10 and a standard error of:

$$SE = \frac{SD}{\sqrt{n}} = \frac{0.50}{\sqrt{30}} = 0.09$$

The Z-score for the workers will be:

$$Z = \frac{6.50 - 8.10}{0.09} = -17.77$$

This is an extremely low Z-score and cannot be looked up on the tables. Thus we can only conclude that the probability of getting a sample of 30 workers with a mean of £6.50 or less is extremely small. It is likely that the workers are genuinely being underpaid.

8.3 Samples of size 100 will be normally distributed with a mean of 29.8 years and a standard error of:

$$SE = \frac{SD}{\sqrt{n}} = \frac{4.6}{\sqrt{100}} = 0.46$$

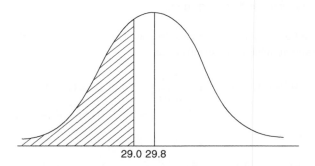

29.0 29.8

Figure A8.3   **Probability of a sample of 100 women with a mean age at the time of giving birth of 29.0 or lower**

We are interested in the probability of getting a sample of 100 women with a mean age at the time of the birth of 29.0 or lower, as in Figure A8.3.

$$Z = \frac{29.0 - 29.8}{0.46} = -1.73$$

The area between the mean and Z = –1.73 is 0.4582.
Area to the left of Z = –1.73 is: 0.5 – 0.4582 = 0.0418.
The chance of getting a sample of 100 women with a mean age at birth of 29.0 or less is quite low (nearly 5%). So we can conclude that the lower sample mean could have occurred due to chance (sampling variation), but that this is not very likely. Therefore, it is likely that the mean age at birth for women in the deprived area is different from the average.

Chapter 9

9.1 (a) It appears that the mean age at which respondents think people generally start being described as old varies by the age of the respondents at the time of being questioned.

(b) Younger age group:

$$95\%\,CI = Sample\,mean \pm \left(1.96 \times \frac{SD}{\sqrt{n}}\right) = 53.15 \pm \left(1.96 \times \frac{11.25}{\sqrt{200}}\right)$$
$$= 53.15 \pm 1.56 = (51.59, 54.71)$$

Older age group:

$$95\%\,CI = Sample\,mean \pm \left(1.96 \times \frac{SD}{\sqrt{n}}\right) = 66 \pm \left(1.96 \times \frac{9.26}{\sqrt{200}}\right)$$
$$= 66 \pm 1.28 = (64.72, 67.28)$$

The confidence intervals for the younger and older age group do not overlap. We can conclude that there is a real difference in the age at which respondents think people generally start being described as old between the younger and older age group.

9.2 (a) The best estimate for the 'death penalty rating' is the sample mean of 3. The standard deviation is given as 5.5.

$$95\%\,CI = 3 \pm \left(1.96 \times \frac{5.5}{\sqrt{40}}\right) = 3 \pm (1.96 \times 0.87)$$
$$= 3 \pm 1.70 = (1.30, 4.70)$$

(b)

$$99\%\,CI = 3 \pm \left(2.575 \times \frac{5.5}{\sqrt{40}}\right) = 3 \pm (2.575 \times 0.87)$$
$$= 3 \pm 2.24 = (0.76, 5.24)$$

This is wider than the 95% confidence interval. If we want to be more confident that the population mean lies within the interval, the interval must be widened.

(c) With a sample of 100 judges:

$$95\%\,CI = 3 \pm \left(1.96 \times \frac{5.5}{\sqrt{100}}\right) = 3 \pm (1.96 \times 0.55)$$
$$= 3 \pm 1.08 = (1.92, 4.08)$$

With a larger sample size, the confidence interval is narrower. This is because a larger sample is more representative of the population and will provide a more precise estimate of the population mean.

9.3 Sample mean = £118,963.63

Sample standard deviation = £67,465.23

Best estimate of mean price is £118,963.63

$$95\% \, CI = £118,963.63 \pm \left(1.96 \times \frac{£67,465.23}{\sqrt{11}}\right)$$

$$= £118,963.63 \pm (1.96 \times £20,341.53)$$

$$= £118,963.63 \pm £39,869.40 = (£79,094.23, £158,833.03)$$

The sample of prices may not be particularly representative because the sample size is rather small and, if the sample is drawn from one estate agent only, we cannot be sure that the prices will be similar in all estate agents.

Chapter 10

10.1 (a) *Males*

Sample proportion $= \dfrac{21}{50} = 0.42$.

Best estimate of population proportion = 0.42.

$$95\% \, CI = 0.42 \pm \left(1.96\sqrt{\frac{0.42 \times (1 - 0.42)}{50}}\right) = (0.2832, 0.5568)$$

Females

Sample proportion $= \dfrac{35}{50} = 0.7$.

Best estimate of population proportion = 0.7

$$95\% \, CI = 0.7 \pm \left(1.96\sqrt{\frac{0.7 \times (1 - 0.7)}{50}}\right) = (0.5730, 0.8270)$$

Although the confidence intervals are quite wide, they do not overlap, so we can say that male students are definitely less likely than female students to believe that one-night stands are usually not a good idea.

(b) *Males*

Sample proportion $= \dfrac{65}{150} = 0.4333$.

Best estimate of population proportion = 0.4333.

$$95\% \, CI = 0.4333 \pm \left(1.96\sqrt{\frac{0.4333 \times (1 - 0.4333)}{150}}\right) = (0.3540, 0.5126)$$

Females

Sample proportion $= \dfrac{110}{150} = 0.7333$.

Best estimate of population proportion = 0.7333.

$$95\% \, CI = 0.7333 \pm \left(1.96\sqrt{\frac{0.7333 \times (1 - 0.7333)}{150}}\right) = (0.6625, 0.8041)$$

The confidence intervals for males and females do not overlap, so there is a significant gender difference in attitudes.

10.2 Proportions in samples of size 100 will normally be distributed with:

Mean proportion = 0.2

$$\text{Standard error} = \sqrt{\frac{\Pi(1-\Pi)}{n}} = \sqrt{\frac{0.2 \times 0.8}{100}} = 0.04$$

Note that for these questions the percentages must be converted into proportions.

(a) See Figure A10.1.

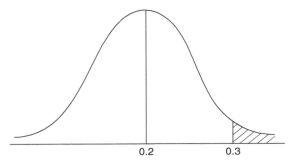

0.2   0.3

Figure A10.1 Probability of 30 or more people declining the offer of an HIV test

$$Z = \frac{0.30 - 0.20}{0.04} = 2.50$$

Area between mean and Z = 2.50 is 0.4938.
Area to the right of Z = 2.50 is: 0.5 – 0.4938 = 0.0062.
The chance that 30 or more people in the sample will decline the offer of an HIV test is 0.0062 (a very low chance).

(b) See Figure A10.2.

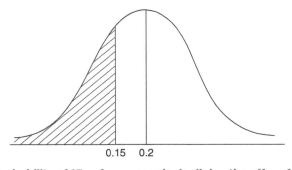

0.15   0.2

Figure A10.2 Probability of 15 or fewer people declining the offer of an HIV test

$$Z = \frac{0.15 - 0.20}{0.04} = -1.25$$

Area between mean and Z = –1.25 is 0.3944.
Area to the left of Z = –1.25 is: 0.5 – 0.3944 = 0.1056.
The chance that 15 or fewer people in the sample will decline the offer of an HIV test is 0.1056.

(c)   See Figure A10.3.

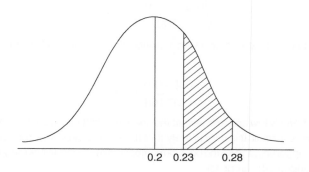

Figure A10.3   **Probability of between 23 and 28 people declining the offer of an HIV test**

For 23:

$$Z = \frac{0.23 - 0.20}{0.04} = 0.75$$

Area between mean and Z = 0.75 is 0.2734.
For 28:

$$Z = \frac{0.28 - 0.20}{0.04} = 2.00$$

Area between mean and Z = 2.00 is 0.4772.
The two areas are both above the mean, so the smaller area is subtracted from the larger area:
0.4772 – 0.2734 = 0.2038.
The chance that between 23 and 28 people will decline the offer of an HIV test is 0.2038.

10.3 (a)   Sample proportion supporting proposal = $\frac{80}{200} = 0.4$.

The best estimate of the population proportion is 0.4.

$$95\% \, CI = 0.4 \pm \left(1.96 \sqrt{\frac{0.4 \times (1 - 0.4)}{200}} \right) = (0.3321, 0.4679)$$

(b)

$$99\% \, CI = 0.4 \pm \left(2.575 \sqrt{\frac{0.4 \times (1 - 0.4)}{200}} \right) = (0.3108, 0.4892)$$

The 99% confidence interval is wider than the 95% confidence interval because if we want to be more certain that the population proportion lies within the range, the range must be wider.

(c) If the true population proportion is 0.6, sample proportions of samples of 200 people will be normally distributed with a mean of 0.6 and a standard error of:

$$SE = \sqrt{\frac{0.6 \times (1 - 0.6)}{200}} = 0.03464$$

What is the chance of getting a sample with a proportion of 0.4 or less from this distribution? See Figure A10.4.

$$Z = \frac{0.4 - 0.6}{0.03464} = -5.77$$

This Z-score is extremely low and tells us that the chance of getting a sample of 200 with a proportion of 0.4 is extremely small. It is very unlikely that the sample did come from a population with a proportion of 0.6 supporting, and so the council member is almost certainly wrong in saying that 60% of the population support the proposal.

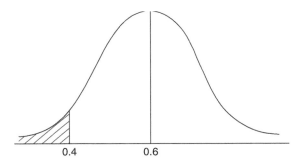

Figure A10.4 **Probability of a sample showing 40% support if there really is 60% support for a proposal**

Chapter 11

11.1 This is a two-sided test for a mean. Your hypotheses are:

H_0: $\mu = 8.2$ (alcohol consumption among 14-year-girls from the local school who had drunk in the last week is the same as the national average)

H_A: $\mu \neq 8.2$ (alcohol consumption among 14-year-girls from the local school who had drunk in the last week is different to the national average)
   Sample mean = 10

Test statistic:

$$Z = \frac{10 - 8.2}{0.8 / \sqrt{34}} = 13.12$$

The critical values are –1.96 and 1.96. The test statistic 13.12 is greater than 1.96 and so lies in the critical region. We can reject the null hypothesis and conclude that there is evidence that alcohol consumption among girls in the sample is significantly different from the national average.

11.2 The question asks for a one-sided test. So the hypotheses are:

H_0: μ = £30,386 (social workers in the sample earn the same as the national average)
H_A: μ < £30,386 (social workers in the sample earn less than the national average)
    Sample mean = £26,555.
    Test statistic:

$$Z = \frac{26,555 - 30,386}{6321 / \sqrt{11}} = -2.01$$

For a one-sided test at the 5% level, the critical value is –1.645. The test statistic falls in the critical region to the left of –1.645, and so we reject the null hypothesis and conclude that the social workers' claim is justified.

11.3 This is a two-sided test for a proportion. The hypotheses are:

H_0: Π = 0.638 (the proportion of people believing that crime prevention is an important issue has not changed)
H_A: Π ≠ 0.638 (the proportion of people believing that crime prevention is an important issue has changed over time)

Sample proportion = $\frac{204}{350} = 0.5829$

Test statistic: Z = $Z = \frac{p - \Pi}{\sqrt{\Pi(1-\Pi)/n}} = \frac{0.5829 - 0.638}{\sqrt{0.638 \times (1 - 0.638)/350}} = -2.14$

The critical values for a two-sided test at the 5% level are –1.96 and 1.96. The test statistic –2.14 is less than –1.96 and so the null hypothesis can be rejected. We conclude that there is evidence to suggest that the proportion of people believing that crime prevention is an important issue has changed significantly over time.

11.4. The question asks for a two-sided test. The hypotheses are:

H_0: μ = £12.94 (the hourly pay rate of men of Pakistani origin in Britain is the same as the hourly pay rate for white men)
H_A: μ ≠ £12.94 (the hourly pay rate of men of Pakistani origin in Britain is different from the hourly pay rate for white men)
    Sample mean = £9.99.

249

Test statistic:

$$Z = \frac{9.99 - 12.94}{2.12 / \sqrt{55}} = -10.32$$

(a) The critical values for a two-sided test at the 5% level are –1.96 and 1.96. The test statistic –10.32 lies well below –1.96, so the null hypothesis can be rejected. It appears that there is evidence that the hourly pay rate for men of Pakistani origin is different to the rate for white men.

(b) For a test at the 1% level, the critical values would be –2.575 and 2.575. The test statistic is still lower than –2.575, so the conclusion remains the same.

11.5 As we do not know what the outcome might be, a two-sided test will be used. The hypotheses are:

H_0: Π = 0.4 (the response rate on a postal survey about an issue of local concern is 40%)
H_A: Π ≠ 0.4 (the response rate on a postal survey about an issue of local concern is different from 40%)

Sample proportion $= \frac{18}{60} = 0.3$

Test statistic: $Z = \frac{0.3 - 0.4}{\sqrt{0.4 \times (1 - 0.4)/60}} = -1.58$

The critical values are –1.96 and 1.96, so the test statistic –1.58 does not lie in the critical region. The null hypothesis must be accepted. There is insufficient evidence at the 5% level to suggest that the researcher's claim is incorrect, so we must assume that it is correct from the evidence we have.

11.6. This is a two-sided test for a proportion. The hypotheses are:

H_0: Π = 0.63 (the proportion of adult smokers at the health centre who want to give up smoking is the same as the national proportion)
H_A: Π ≠ 0.63 (the proportion of adult smokers at the health centre who want to give up smoking is different to the national proportion)

Sample proportion $= \frac{89}{124} = 0.7177$

Test statistic: $Z = \frac{0.7177 - 0.63}{\sqrt{0.63(1 - 0.63)/40}} = 1.15$

For a two-sided test, the critical values will be –1.96 and 1.96. The test statistic does not lie in the critical region and so the null hypothesis must be accepted. Smokers at the health centre are not significantly different from those in the country as a whole in their desire to give up.

Chapter 12

12.1 This is a two-sided test. The population standard deviation is unknown and the sample size is less than 30, so a *t* test should be used.

H_0: μ = £297 (the weekly household expenditure of lone mothers in your area is the same as the national average)
H_A: $\mu \neq$ £297 (the weekly household expenditure of lone mothers in your area is different from the national average)
   Test statistic:

$$t = \frac{274 - 297}{33 / \sqrt{16}} = -2.79$$

From a sample of 16, there are 15 degrees of freedom. For a two-sided test at the 5% level, t_{15} = 2.1314 (look down the t_{df} (0.025) column in Table A1.2 because there is 2.5% in each tail). The critical values are –2.1314 and 2.1314. The test statistic lies outside this range, so the null hypothesis most be rejected. The household expenditure of lone mothers in your area is different from the national average.

12.2 The question asks for a one-tailed test, and a t test is used because the standard deviation of the population is not known and the sample size is less than 30.

H_0: μ = £1.80 (the children are receiving the average amount of pocket money)
H_A: $\mu <$ £1.80 (the children are receiving less pocket money than average)

Test statistic: $t = \dfrac{1.56 - 1.80}{0.45 / \sqrt{25}} = -2.67$

With a sample of 25, there will be 24 degrees of freedom. For a one-sided test at the 5% level there is 5% in one tail of the distribution, so we look down the t_{df}(0.05) column in Table A1.2 to obtain t_{24} = 1.7109. The critical value will be –1.7109 because we are trying to find out whether the children are getting significantly less pocket money than the mean.
The test statistic –2.67 is less than –1.7109, so the null hypothesis can be rejected. The children are receiving significantly less pocket money than the mean.

12.3 For this question the hypotheses are:

H_0: Π_m = 0.5 (the charges for weekly childcare in various regions of England come from a population where half the observations are above £133.17; charges in other regions are the same as in London)

H_A: $\Pi_m \neq$ 0.5 (the charges for weekly childcare in various regions of England do not come from a population where half the observations are above £133.17; charges in other regions are not the same as in London)
There are 8 negative signs (observations below £133.17) and 0 positive signs (observations above £133.17). Thus the sample proportion is:

$$p_m = \frac{\text{number of plus signs}}{\text{total}} = \frac{0}{8} = 0$$

$$\text{Test statistic}: Z = \frac{p_m - \Pi_m}{\sqrt{\Pi_m(1-\Pi_m)/n}} = \frac{0 - 0.5}{\sqrt{0.5(1-0.5)/8}} = -2.83$$

The critical values are –1.96 and 1.96. The test statistic is outside the range of the critical values, so the null hypothesis can be rejected.

12.4 For this question the hypotheses are:

H_0: $\Pi_m = 0.5$ (sponsorship amounts come from a population with half the observations above £63.20; the amount raised per rider has not changed since last year)

H_A: $\Pi_m \neq 0.5$ (sponsorship amounts do not come from a population with half the observations above £63.20; the amount raised per rider has changed since last year)
There are 12 positive signs (amounts over £63.20) in the sample. The sample proportion is:

$$p_m = \frac{12}{20} = 0.6$$

$$\text{Test statistic}: Z = \frac{p_m - \Pi_m}{\sqrt{\Pi_m(1-\Pi_m)/n}} = \frac{0.6 - 0.5}{\sqrt{0.5(1-0.5)/20}} = 0.89$$

The critical values are –1.96 and 1.96. The test statistic lies within this range, so the null hypothesis must be accepted. There has been no significant change in sponsorship money per rider since last year.

Chapter 13

13.1 (a) The mean first-semester exam marks will be the Y variable because they may depend on the A-level scores.
   (b) The scatter graph (Figure A13.1) does not show a very strong association between A-level scores and first-semester exam results.
   (c) Calculation of the correlation coefficient:

Using the totals from the table, the correlation coefficient is

$$r = \frac{113.48}{\sqrt{265.74 \times 761.14}} = 0.25$$

This indicates a very weak positive correlation between A-level scores and first- semester exam marks among the 18 students.

(a) $r^2 = 0.25^2 = 0.06$ (to two decimal places).

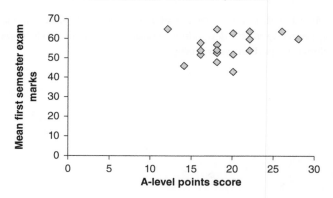

Figure A13.1 A-level points score and mean first-semester exam mark for 18 students (anonymous student records)

X	Y	$(X_i - \bar{X})$	$(Y_i - \bar{Y})$	$(X_i - \bar{X})^2$	$(Y_i - \bar{Y})^2$	$(X_i - \bar{X})(Y_i - \bar{Y})$
18	54	−1.11	−2.22	1.23	4.93	2.46
18	48	−1.11	−8.22	1.23	67.57	9.12
16	52	−3.11	−4.22	9.67	17.81	13.12
18	65	−1.11	8.78	1.23	77.09	−9.75
14	46	−5.11	−10.22	26.11	104.45	52.22
12	65	−7.11	8.78	50.55	77.09	−62.43
20	52	0.89	−4.22	0.79	17.81	−3.76
18	57	−1.11	0.78	1.23	0.61	−0.87
20	63	0.89	6.78	0.79	45.97	6.03
18	53	−1.11	−3.22	1.23	10.37	3.57
26	64	6.89	7.78	47.47	60.53	53.60
16	54	−3.11	−2.22	9.67	4.93	6.90
20	43	0.89	−13.22	0.79	174.77	−11.77
16	58	−3.11	1.78	9.67	3.17	−5.54
28	60	8.89	3.78	79.03	14.29	33.60
22	64	2.89	7.78	8.35	60.53	22.48
22	60	2.89	3.78	8.35	14.29	10.92
22	54	2.89	−2.22	8.35	4.93	−6.42
$\bar{X} = 19.11$	$\bar{Y} = 56.22$		Totals	265.74	761.14	113.48

A-level scores are not a good predictor of first semester exam marks among this group of students; A-level scores only explain 6% of the difference in mean exam marks. Perhaps motivation, ability to work independently or other factors are more important than past achievements.

13.2 (a)   Figure A13.2 shows a positive relationship between life expectancy at birth and the percentage of the population with access to drinking water sources which have been improved to make them safe in the 10 countries. Note that life expectancy is the dependent variable here because life expectancy may depend in part on water quality.

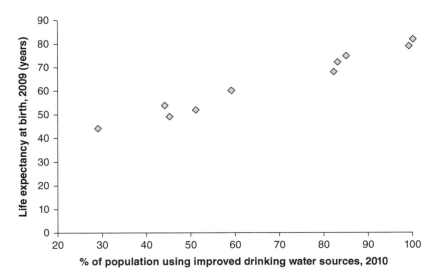

Figure A13.2 **Life expectancy at birth and percentage of population with access to drinking water sources which have been improved to make it safe in 10 countries**

Source: World Health Organization, 2012

(b) Calculation of the correlation coefficient:

X	Y	$(X_i - \bar{X})$	$(Y_i - \bar{Y})$	$(X_i - \bar{X})^2$	$(Y_i - \bar{Y})^2$	$(X_i - \bar{X})(Y_i - \bar{Y})$
100	82	32.30	18.50	1043.29	342.25	597.55
85	75	17.30	11.50	299.29	132.25	198.95
51	52	−16.70	−11.50	278.89	132.25	192.05
29	44	−38.70	−19.50	1497.69	380.25	754.65
83	72	15.30	8.50	234.09	72.25	130.05
44	54	−23.70	−9.50	561.69	90.25	225.15
59	60	−8.70	−3.50	75.69	12.25	30.45
45	49	−22.70	−14.50	515.29	210.25	329.15
82	68	14.30	4.50	204.49	20.25	64.35
99	79	31.30	15.50	979.69	240.25	485.15
$\bar{X} = 67.70$	$\bar{Y} = 63.50$		Total	5690.10	1632.50	3007.50

Using the totals from the table, the correlation coefficient is

$$r = \frac{3007.50}{\sqrt{5690.10 \times 1632.50}} = 0.99$$

A correlation of 0.99 indicates a strong positive correlation: countries where a high percentage of the population has access to drinking water sources which have been improved to make them safe are likely to have high life expectancy at birth.

(c) Here you should have already calculated the values to put into the equations during the calculation in (b) to find the correlation coefficient.

$$b = \frac{\sum\left[(X_i - \bar{X})(Y_i - \bar{Y})\right]}{\sum(X_i - \bar{X})^2} = \frac{3007.50}{5690.10} = 0.53$$

$$a = \bar{Y} - b\bar{X} = 63.50 - (0.53 \times 67.70) = 63.50 - 35.88 = 27.62$$

(d) The regression line is Y = a + bX.

Life expectancy at birth is equal to 27.62 plus 0.53 times the percentage of those with access to drinking water sources which have been improved to make them safe.

(e) If nobody in a population had access to drinking water sources which had been improved to make them safe, life expectancy would be predicted to be:

27.62 + (0.53 × 0) = 27.62 years

This is a possible figure for life expectancy at birth, but it may be unrealistic to expect that nobody in a country has access to drinking water sources which have been improved to make them safe. In any case, the value of 0% with access to drinking water sources which have been improved to make it safe is outside the range of the data, so we should not predict life expectancy from it.

(f) (i) If 40% of the population have access to drinking water sources which have been improved to make them safe, we predict a life expectancy of
Y = 27.62 + (0.53 × 40) = 48.82 years

(ii) If 52% of the population have access to drinking water sources which have been improved to make them safe, we predict a life expectancy of
Y = 27.62 + (0.53 × 52) = 55.18 years

(iii) If 90% of the population with access to drinking water sources which have been improved to make them safe, we predict a life expectancy of
Y = 27.62 + (0.53 × 90) = 75.32 years

Chapter 14

14.1 (a) See Table A14.1. It appears that a larger proportion of people agreed that a mother should be able to work on a full- or part-time basis when there is a child under school age in 2012 than in 1989.

(b) The hypotheses are:

H_0: there is no association between year and attitudes towards mothers' employment

H_A: there is an association between year and attitudes towards mothers' employment

Table A14.1 Attitudes to mothers' employment when there is a child under school age in 1989 and 2012

Number agreed a mother should	Year		
	1989	2012	Total
work full-time	25	47	72
work part-time	331	410	741
stay at home	678	314	992
Total	1034	771	1805

Source: Park et al., 2013

Table A14.2 Expected values for question 14.1(b)

Number agreed a mother should	Year		
	1989	2012	Total
work full-time	41.25	30.75	72
work part-time	424.48	316.52	741
stay at home	568.27	423.73	992
Total	1034.00	771.00	1805

Source: Park et al., 2013

The expected values are shown in Table A14.2. The test statistic is calculated as follows:

O	E	$\dfrac{(O-E)^2}{E}$
25	41.25	6.40
331	424.48	20.59
678	568.27	21.19
47	30.75	8.59
410	316.52	27.61
314	423.73	24.42
	Total	108.80

$$\chi^2 = 108.80$$

There are two degrees of freedom and so, at the 5% level, $\chi^2 (0.05) = 5.9915$.

The test statistic 108.80 is bigger than 5.9915, so it lies in the critical region. We must reject the null hypothesis and conclude that there is a significant association at the 5% level between year and attitudes towards mothers' employment.

14.2 (a)   The hypotheses are:

H_0: there is no association between the year and interest in politics
H_A: there is an association between the year and interest in politics

Table A14.3   Expected values for question 14.2(a)

Interest in politics	Year		Total
	1986	2012	
Great deal/quite a lot	492.00	353.02	845
Some	484.41	347.59	832
Not much/none at all	556.60	399.39	956
Total	1533	1100	2633

Source: Park et al., 2013

The expected values are shown in Table A14.3 and the test statistic is calculated as follows:

O	E	$\dfrac{(O-E)^2}{E}$
449	492.00	3.76
480	484.41	0.04
604	556.60	4.04
396	353.02	5.23
352	347.59	0.06
352	399.39	5.62
	Total	18.75

$$\chi^2 = 18.75$$

There are two degrees of freedom and so, at the 5% level, $\chi^2(0.05) = 5.9915$.
The test statistic is well above the critical value (18.75 > 5.9915) and so the null hypothesis can be rejected. There is an association between the year and interest in politics.

(b)   If the test were carried out at the 1% level, the critical value would be 9.2103. The test statistic 18.75 is greater than this, so the null hypothesis would still be rejected. There is a significant association at the 1% level between the year and interest in politics.

(c)   The residuals are shown in Table A14.4. All six of the residuals are significant because they are greater than 2 or less than –2. We can conclude that those asked about levels of interest in politics in 2012 are significantly more likely to have a great deal/quite a lot or some interest in politics than 1986 (because the residuals are positive) and significantly less likely to have not much or no interest at all in politics in 2012 (because the residual is negative) than people in 1986.

Table A14.4   Residuals for question 14.2(c)

	Year	
	1986	**2012**
Great deal/quite a lot	−2.53	2.93
Some	−4.54	5.25
Not much/none at all	4.60	−4.94

14.3 (a)   The expected values are shown in Table A14.5. Three of the expected values are below 5. This is a problem if we want to carry out a χ^2 test.

Table A14.5   Expected values for question 14.3 (a)

Age	Do you think that euthanasia should be legalised in Britain?			Total
	Yes	**No**	**Don't know**	
18–34	20.42	16.25	13.33	50
35–54	16.74	13.33	10.93	41
55–64	7.76	6.18	5.07	19
65+	4.08	3.25	2.67	10
Total	49	39	32	120

(b)   The categories 55–64 and 65+ are combined or collapsed and a new table has been calculated (Table A14.6) along with new expected values (Table A14.7).

Table A14.6   Attitudes towards legalising euthanasia in Britain by age after collapsing the 'age variable' categories

Age	Do you think that euthanasia should be legalised in Britain?			Total
	Yes	**No**	**Don't know**	
18–34	25	10	15	50
35–54	16	15	10	41
55+	8	14	7	29
Total	49	39	32	120

Source: Hypothetical data

(c)   The hypotheses are:

H_0: there is no association between age and whether people believe euthanasia should be legalised in Britain

H_A: there is an association between age and whether people believe euthanasia should be legalised in Britain

There are $2 \times 2 = 4$ degrees of freedom, so the critical value at the 5% level will be 9.4877. The test statistic is smaller than the critical value, so the null hypothesis would be accepted. There is no association between age and whether people believe euthanasia should be legalised in Britain.

Table A14.7   Expected values for question 14.3 (b)

| Age | Do you think that euthanasia should be legalised in Britain? | | | Total |
	Yes	No	Don't know	
18–34	20.42	16.25	13.33	50
35–54	16.74	13.33	10.93	41
55+	11.84	9.43	7.73	29
Total	49	39	32	120

Source: Hypothetical data

The test statistic is calculated as follows:

O	E	$\dfrac{(O-E)^2}{E}$
25	20.42	1.03
16	16.74	0.03
8	11.84	1.25
10	16.25	2.40
15	13.33	0.21
14	9.43	2.21
15	13.33	0.21
10	10.93	0.84
7	7.73	0.07

$$\chi^2 = 8.25$$

APPENDIX 3

ALGEBRA AND MATHEMATICAL NOTATION EXPLAINED

If you struggled with algebra and mathematical notation at school, or indeed if school itself is a very distant memory, this appendix is for you. You could either check through it before working your way through the book to get your algebra and mathematical notation skills up to scratch, or use it as a reference guide if you get stuck later on.

Algebra

We have deliberately not called this section 'Basic algebra' or 'Simple algebra' because you may feel initially that it is anything but simple! However, algebra is really all about following a set of rules, so provided that you understand and follow the rules, you cannot go far wrong. There is nothing mysterious about it!

There are two main things you need to know: what order to do things in, and how to move something from one side of an equation to the other.

What order to do things in!

Equations are often full of +, −, × and ÷ signs, along with brackets. Where do you start?

The main thing to remember is to work out anything inside brackets first. For example:

$$(10 + 8) \div 2$$

Here you should work out the 10 + 8 in brackets first and then divide by 2 to get an answer of 9:

$$(10 + 8) \div 2 = 18 \div 2 = 9$$

However, if the same equation was written with the brackets in a different place, the answer would be different! In the following example, the 8 ÷ 2 in brackets must be calculated first, before adding onto the 10:

$$10 + (8 \div 2) = 10 + 4 = 14$$

Brackets can clearly make quite a big difference!

Sometimes there will not be any brackets in an equation, where perhaps there should be. In such cases, × and ÷ operations should always be carried out before + and – operations.

If our equation was written with no brackets, as follows, the $8 \div 2$ should be calculated before adding the 10, because dividing has priority over adding:

$$10 + 8 \div 2 = 10 + 4 = 14$$

Moving things from one side of an equation to the other

The most common type of equation that you will meet will be something like this:

$$X + 2 = 5 \rightarrow X = ?$$

The aim is to solve the equation by working out what the value of X is. To do this you need to get X on its own on one side of the equation and all the other numbers on the other side. But how can we move something over to the other side?

The answer is to move it and change the +, –, × or ÷ to its opposite:

+ changes to –

– changes to +

× changes to ÷

÷ changes to ×

To solve $X + 2 = 5$, the 2 needs to be moved to the other side of the = sign, to leave the X on its own. As the 2 is added to the left-hand side, it can be subtracted from the right-hand side to solve the equation:

$$X + 2 = 5 \quad \rightarrow \quad X = 5 - 2 \quad \rightarrow \quad X = 3$$

The following are some more examples of equations solved using this method:

$$X - 3 = 8 \quad \rightarrow \quad X = 8 + 3 \quad \rightarrow \quad X = 11$$
$$X \div 4 = 10 \quad \rightarrow \quad X = 10 \times 4 \quad \rightarrow \quad X = 40$$
$$X \times 2 = 30 \quad \rightarrow \quad X = 30 \div 2 \quad \rightarrow \quad X = 15$$

More complex equations may involve the use of brackets and negative numbers. The following examples show in detail how four equations are solved. The workings are given on the left, with explanations on the right.

Example 1

$\dfrac{X+5}{10}=2.5$	To solve the equation, we must get X on its own
$X+5=2.5\times10$	The whole left-hand side was divided by 10. The 10 has been moved over to the other side and is now multiplied (remember $\times$ is the opposite of $\div$)
$X+5=25$	The right-hand side is worked out
$X=25-5$	The $+5$ is moved to the other side and becomes -5.
$X=20$	The equation is solved!

Example 2

$X-(2\times3)=10$	
$X-(6)=10$	The section in brackets was worked out first.
$X=10+6$	The 6 is subtracted from the left-hand side and so can be moved to the right-hand side and added.
$X=16$	The equation is solved.

Example 3

$\dfrac{X}{5}=6+(-2)$	
$\dfrac{X}{5}=4$	The right-hand side has been calculated. The $+$ and $-$ signs combine to make a $-$, and $6-2=4$.
$X=4\times5$	The left-hand side was divided by 5, so the 5 is moved to the right-hand side and is multiplied.
$X=20$	The equation is solved.

Example 4

$3X+2=11$	
$3X=11-2$	To remove the $+2$ from the left-hand side it is subtracted from the right-hand side.
$3X=9$	The right-hand side is worked out.
$X=9\div3$	$3X$ means 3 times X. To get rid of the $\times3$, it must be moved to the other side and changed to $\div3$.
$X=3$	The equation is solved.

Mathematical notation explained

This book avoids using notation as much as possible. However, you will need to understand what various bits of notation mean when you see them elsewhere.

Believe it or not, mathematical jargon is designed to make life easier for you rather than harder! Using letters is much quicker than writing down lots of numbers, and fortunately there are only a few things you need to know about at this stage.

Often we use letters as shorthand for variables. For example, we might be investigating population growth and infant mortality in eight countries. We could call the two variables X and Y for speed, so:

Population growth = X

Infant mortality = Y

Of the eight countries, the first is Tanzania and the second Kenya. In shorthand therefore:

Population growth in Tanzania = X_1

Infant mortality in Kenya = Y_2

So the observations will be called $X_1, X_2, ..., X_8$ and $Y_1, Y_2, ..., Y_8$.

However, this is still quite a long-winded way of writing things. To shorten what we need to write even further, we can use the letter i, where i means any of the countries. The observation for the ith country will be X_i. This is particularly useful when we are adding up a set of observations, for example to calculate a mean or standard deviation.

In Chapter 4 we met Σ, the Greek capital letter sigma, which means 'sum of'. In that chapter, we simply wrote ΣX to indicate 'sum of all the values of X'. However, in most texts you will read, it is written more accurately as follows. To indicate 'sum of values for population growth in the eight countries', we should write:

$$\sum_{i=1}^{8} X_i$$

You can see that X_i has been used as the proper way of writing 'any value of X', and the numbers above and below the Σ indicate 'observations 1 to 8'.

Similarly if we wanted to write 'sum of values for infant mortality for countries 1 to 4', it would be:

$$\sum_{i=1}^{4} Y_i$$

Check that you understand what you have just read by answering the following questions.

Questions: Notation

The following are the ages of four women when they were married for the first time:

Woman	Age	Notation
Pauline	32	X_1
Ramona	21	X_2
Amy	24	X_3
Katrina	27	X_4

What do the following expressions mean? Work out what number they are equivalent to, using the data in the table.

(a) $\displaystyle\sum_{i=1}^{4} X_i$

(b) $\displaystyle\left(\sum_{i=1}^{3} X_i\right)^2$

(c) $\displaystyle\left(\sum_{i=1}^{2} X_i\right)^2 + \left(\sum_{i=3}^{4} X_i\right)^2$

Answers: Notation

(a) The sum of observations 1 to 4:

$$32 + 21 + 24 + 27 = 104$$

(b) The sum of observations 1 to 3, all squared. Note that everything inside the bracket is done first, before squaring:

$$(32 + 21 + 24)^2 = 77^2 = 5929$$

(c) Observations 1 plus 2, squared, added to observations 3 plus 4, squared. Again, the sections in brackets must be worked out first:

$$(32 + 21)^2 + (24 + 27)^2 = 53^2 + 51^2 = 2809 + 2601 = 5410$$

If you can cope with these questions, you should have no problems with understanding notation.

Hopefully this appendix will have given you some hints on coping with the maths involved with statistics. If these kinds of things are still causing you great problems, try looking at a GCSE maths textbook.

GLOSSARY

Abridged frequency table – A frequency table in which the data are grouped.

Alternative hypothesis – In hypothesis testing, this is the proposition that there will be a relationship.

Array – When data are placed in a numerical order.

Attributes – The particular characteristics of a variable.

Back-to-back stem and leaf plot – See **stem and leaf plot**.

Bar chart – Visual representation of data used to describe nominal and ordinal variables. Bars are used to represent the count, percentage or proportion of each category of the variable.

Bimodal – A distribution with two modes.

Box plot – Graph used to summarise a variable's distribution. The graph displays the median, upper and lower quartile, and any outliers.

Case – This represents those who are the object of the study. So each case may represent an individual, organisation or community.

Categorical variable – A variable whose attributes have been categorised. Such variables have no numerical value. Nominal and ordinal variables are categorical.

Causal – A relationship between two variables in which the independent variable is believed to influence the dependent variable.

Cell – The intersection of a row and a column in a cross-tabulation or contingency table.

Central tendency – Statistics used to determine what is average or typical in the distribution of data. There are three different ways of doing this: the mean, median and mode.

Centiles – See **percentiles**.

Central limit theorem – Theorem which states that when the sums or means of large samples of random samples are plotted, the shape will be a normal distribution, a bell-shaped curve.

Chi-square test – A test used to determine the statistical significance of a hypothesis involving categorical variables by providing a means of determining whether a set of observed frequencies deviate significantly from a set of expected frequencies.

Cluster sampling – A random sampling technique which involves dividing the population into clusters, randomly sampling clusters and measuring all units within sampled clusters.

Code book – A document which gives information about each variable in a data file, such as name and type, along with the units of measurement or categories.

Column percentages – Percentages in a table which add up to 100% vertically.

Complete frequency distribution – The process of listing every value and the 'frequency' or number of times a case occurs.

Confidence interval – Range of values that is likely to contain the true value of the population parameter.

95% confidence interval – You can make estimates for any level of confidence you like. However, most often people use the two limits which we are 95% certain the value will lie between: this is called a 95% confidence interval.

Contingency table – A frequency table providing the frequencies in all of the categories of two or more categorical variables tabulated together.

Continuous variable – A variable where an observation may take any value on a continuous scale. Interval and ratio variables are types of continuous variables.

Convenience sample – A non-probability sampling technique which involves the selection of a sample based on those available to the researcher as a result of its accessibility.

Correlation – Measures the association between two continuous variables. Correlations can be positive or negative. A correlation is positive when the values increase together and negative when one value decreases as the other increases.

Correlation coefficient – This measures the strength and direction of the relationship between two variables. The most common correlation coefficient is Pearson's r.

Counts – This refers to the number of times an attribute occurs in a cell.

Critical region – The region of a test statistic where a null hypothesis is rejected via the results of a hypothesis test.

Critical values – The point(s) on the scale of the test statistic beyond which we reject the null hypothesis.

Cross-tabulation – A table of the joint frequency distributions of two nominal or ordinal variables, sometimes called a contingency table.

Cumulative frequency – Shows the number of data points falling into a category as well as all its preceding categories, in a frequency table.

Cumulative percentage – Shows the percentage of the data points falling into a category as well as all its preceding categories.

Cumulative proportion – Shows the proportion of the data points falling into a category as well as all its preceding categories.

Curvilinear – The data follow a curved pattern.

Data – A collection of observations.

Deductive – An approach to the relationship between theory and research in which the latter is conducted with reference to the hypotheses.

Dependent variable – The variable that is being predicted or caused by the independent variable.

Descriptive statistics – Tools used to describe data and their characteristics.

Dichotomous variable – A variable which has only two categories or levels. For example, if we were looking at gender, we would usually categorise somebody as 'male' or 'female'.

Discrete variable – See **categorical variable**.

Distribution – How the values of a variable are spread.

Expected values – The values which we would expect to appear if variables are completely independent of each other.

Grouping – The process of putting similar data together. This may involve recoding variables.

Hi-lo plot – This shows the estimated mean and the confidence interval for one or more populations.

Histogram – A bar graph used for interval and ratio (continuous) variables.

Hypothesis – Designed to express a relationship between variables that can be tested.

Independent variable – The variable used to try to predict a dependent variable.

Inferential statistics – Statistical tests that are used to draw inferences about differences between groups or relationships between variables within a population.

Interpolate – To estimate the value of something given certain data either side of the point of interest.

Inter-quartile range – The difference between the highest and lowest values in a distribution of values when the highest and lowest 25% of values have been removed; or the difference between the **upper quartile** and **lower quartile**.

Interval estimate – A range within which we can state how confident we are that the real population statistic of interest lies within the range.

Interval variable – A continuous variable where categories may be ranked or ordered and the distance between them is clearly defined. It has an arbitrary zero point.

Leaves – The numbers to the right of the line in a stem and leaf plot.

Linear – A relationship between two variables in which values on one show an even increase or decrease as values on the other change, seen as a straight line on a graph.

Line graph – A diagram in which lines are used to indicate the frequency of a variable.

Lower quartile – The lower quartile is the 25th percentile, so 25% of observations will lie at or below the lower quartile.

Mean – Sometimes referred to as the average, it is the sum of all of the values in a dataset, divided by the number of values.

Median – The middle observation is known as the median. Half of the observations will lie above the median and half below.

Mode – The mode is the most frequently occurring value in the data.

Normal distribution – The tendency for data to cluster around a central point in a predictable and symmetric manner.

Normal table – Used to find the area between the mean and any Z-score. The normal table gives the proportion of the area of the curve between the mean and a certain positive value of Z.

Nominal variable – A variable which has categories or attributes that are not ordered and have no inherent numerical quality.

Non-parametric test – A statistical test which needs fewer assumptions about the distribution of values in a sample than a parametric test.

Non-sampling error – These are errors not connected with the sampling method. For instance, interviewers can unconsciously bias the results or record question responses incorrectly.

Null hypothesis – In hypothesis testing, this is the proposition that there will not be a relationship.

Observations – The responses that you get from each case.

Observed values – The values that are actually obtained when conducting research.

One-sided test – For a one-sided test, the null hypothesis is rejected if the test statistic is larger than the critical value.

Opinion poll – An assessment of public opinion by questioning a sample as the basis for forecasting the results on a larger scale.

Ordinal variable – A variable where values can be ordered or ranked moving from greater to smaller values (or vice versa). While the values are in a particular order they have no exact distance between them. The distance between the points on the scale is not clear and continuous.

Outlier – An observation which falls outside a typical pattern.

Parallel line plot – The use of two lines plotted on the same scale to enable a straight comparison.

Pearson product moment correlation coefficient – A measure of the linear correlation (dependence) between two variables.

Percentage – A number or ratio normally expressed as a score out of 100. It is similar to a proportion but a proportion can take any value from 0 to 1 inclusive, while a percentage usually takes a value from 0% to 100% (although percentage changes greater than 100% are possible).

Percentiles – These divide a set of data into hundredths (100 equal parts).

Pie chart – A graph used to describe the distribution of nominal or ordinal data. A pie chart presents the categories of data as parts of a circle or 'slices of a pie'.

Point estimate – The use of a sample statistic such as the mean to estimate the population parameter of interest instead of a whole range of values.

Population – The entire set of people or events to which the research hypothesis is presumed to apply.

Predict – To estimate that something will happen.

Probability – The likelihood of an occurrence.

Proportion – A proportion is very similar to a percentage, the main difference being that a proportion can take any value from 0 to 1 inclusive, while a percentage usually takes a value from 0% to 100% (although percentage changes greater than 100% are possible).

Purposive sampling – This involves sampling with a purpose in mind. It involves selecting a sample based on knowledge of a population, its subgroups, and the purpose of the study, selecting people who would be 'good' informants.

Quota sampling – Participants are selected non-randomly according to some fixed quota. It is used to provide a sample that reflects a population in terms of relative proportions of people in different categories such as gender, age or ethnicity.

Range – The difference between the smallest and largest values in a distribution.

Ratio variable – A continuous variable where categories may be ranked or ordered and the distance between them is clearly defined. It has a true zero point.

Regression – A technique designed to predict values of a dependent variable from knowledge of the values of one or more independent variables.

Relative frequency – Another term for proportion, calculated by dividing the number of times an event occurs by the total number of times it is carried out.

Residual – The difference between a particular observation and the mean.

Row percentages – Percentages in a table which add up to 100% horizontally.

Sample – A subset of a population.

Sampling error – The extent to which the sample does not reflect the population as a result of differences between the population and the sample generated by random selection of cases.

Sample mean – The mean from a group of observations in a sample. Can be used as an estimate of the population mean.

Sampling variability – The different values which a given function of the data (such as the mean) takes when computed for two or more samples drawn from the same population.

Scaling – Adding or subtracting a constant to the data or multiplying or dividing the data by a constant.

Scatter graph – Graph of the relationship between two continuous variables X and Y.

Significance test – A test of whether the alternative hypothesis achieves the predetermined significance level in order to be accepted in preference to the null hypothesis.

Sign test – The one-sample sign test compares the median of the sample with the population median. This test can be used when the data do not follow the normal distribution.

Simple random sample – A sample in which every member of a population has an equal chance of being selected.

Skewed – When a distribution is skewed most of the data are concentrated at one end of the range.

Snowball sampling – A form of non-probability sampling where you begin by identifying someone who meets the criteria for inclusion in your study. You then make contact and ask them to recommend others they may know who also meet the criteria. This process continues to make up your sample.

Spread – A measure of spread summarises how variable a distribution is, that is, how tightly cases in a frequency distribution cluster around their central tendency (average).

Standard deviation – A measure of spread. The standard deviation is equal to the square root of the variance, the sum of the squared deviations of the observations from their mean which is divided by the number of observations minus one. The larger is the standard deviation, the greater is the spread of the data.

Standard error – The standard error is equal to the standard deviation of the population divided by the square root of the sample size. We call the standard deviation of the sample means the standard error to distinguish it from the ordinary standard deviations of samples or populations.

Standardised variable – A variable where the observations have been converted into Z-scores.

Statistics – A set of tools and techniques used to organise and interpret information.

Stem – The column in a stem and leaf plot to the left of the line.

Stem and leaf plot – A stem and leaf plot is a type of graph that summarises the shape of a set of data. A basic stem and leaf display contains two columns separated by a vertical line. The left column contains the *stems* (the first digit(s)) and the right column contains the *leaves* (usually the last digit). For example, '45' would be split into '4' (stem) and '5' (leaf). In order to compare two different distributions, two stem and leaf plots can be drawn next to each other, sharing the same stem. This is known as a **back-to-back stem and leaf plot**.

Stratified sample – If we know there are specific groups in a population who may be different from each other, we might select the sample to reflect the various strata in the population.

Strong correlation – If all the points on a scatter graph lie in a fairly straight line, we have a strong correlation between the two variables.

Systematic sample – Systematic sampling is similar to simple random sampling. It is a form of random sampling but it involves a system. From the sampling frame, a starting point is identified at random and a selection made at regular intervals.

t **distribution** – Used to find the critical region for tests of sample means when the sample size is small and the population is normally distributed.

t **test** – A significance test which compares the sample mean with the population mean. The *t* test can only be used when the data are normally distributed.

Test statistic – The statistic used to test a null hypothesis. The distribution is known. Common test statistics are the Z-score, the *t*-value and chi-square.

Time series – A set of data which shows changes in a variable over time.

Two-sided tests – For a two-sided test, the null hypothesis is rejected if it falls into either the critical region above the mean value, or the critical region below the mean value.

Upper quartile – The upper quartile is the 75th percentile, so 75% of observations will lie at or below it and 25% at or above it.

Variable – Properties or characteristics of the cases which differ across observations.

Variance – The variance is the standard deviation squared. Omitting the last stage of the standard deviation formula where the square root of the value is taken will give the variance instead of the standard deviation.

Weak correlation – If the points on a scatter graph are spread out but still show some relationship, we have a weak correlation between the two variables.

X axis – On any graph, the horizontal axis is known as the X axis.

Y axis – On any graph, the vertical axis is known as the Y axis.

Z-scores – Standardisation of a value by subtracting the variable's mean from the value and dividing by the variable's standard deviation.

REFERENCES

Acton, C., Millar, R. with Fullarton, D. and Maltby, J. (2009) *SPSS for Social Scientists.* 2nd edn. Basingstoke: Palgrave Macmillan.

Alzheimer's Society (2012) *Dementia: A National Challenge Report.* London: Alzheimer's Society. http://www.alzheimers.org.uk/site/scripts/download_info.php?fileID=1389.

American Psychological Association (2010) *Publication Manual of the American Psychological Association,* 6th edn. Washington, DC: American Psychological Association.

AXA Wealth (2012) *AXA Cost of Living in Retirement Report.* Basingstoke: AXA Wealth. https://www.axa-lifeinvest.co.uk/wps/themes/Portal/ASSETS_UK/Documents/Press%20 Releases/AXA%20Cost%20of%20Living%20in%20Retirement%20Report%20(2).pdf.

Babbie, E., Halley, F. and Ziano, J. (2012) *Adventures in Social Research: Data Analysis Using SPSS,* 8th edn. Thousand Oaks, CA: Pine Forge Press.

Banks, J., Tetlow, G. and Wakefield, M. (2008) *Asset Ownership, Portfolios and Retirement Saving Arrangements: Past Trends and Prospects for the Future.* Consumer Research 74. London: Financial Services Authority. http://www.fsa.gov.uk/pubs/consumer-research/ crpr74.pdf.

Baudrillard, J. (1990) *Cool Memories.* London: Verso.

Becker, H. (1963) *Outsiders: Studies in Sociology of Deviance.* New York: Free Press.

Booth, C. (1886) 'Occupations of the United Kingdom 1801–1881', *Journal of the Statistical Society of London,* 49(2), 314–444.

Booth, C. (1902) *Life and Labour of the People in London,* Vol. 1. London: Macmillan.

Bryman, A. and Cramer, D. (2011) *Quantitative Data Analysis with IBM SPSS 17, 18 and 19: A Guide for Social Scientists.* London: Routledge.

Burdess, N. (2010) *Starting Statistics – A Short Clear Guide.* London: Sage.

Butler, D. and Kavanagh, D. (1992) *The British General Election of 1992.* Basingstoke: Macmillan.

Cribb, J., Joyce, R. and Phillip, D. (2012) *Living Standards, Poverty and Inequality in the UK: 2012.* Institute of Fiscal Studies Commentary C124. London: IFS. http://www.ifs.org.uk/ comms/comm124.pdf.

Daycare Trust and Family and Parenting Institute (2013) *Childcare Costs Survey 2013.* London: Daycare Trust.

Davis, A., Hirsch, D., Smith, N., Beckhelling, J. and Padley, M. (2012) *A Minimum Income Standard for the UK in 2012: Keeping Up in Hard Times.* York: Joseph Rowntree Foundation.

Davis, C. (2013) *SPSS Step by Step: Essentials for Social and Political Science.* Bristol: Policy Press.

Department for Communities and Local Government (2012) *The Troubled Families Programme.* http://www.communities.gov.uk/documents/communities/pdf/2117840.pdf.

Department for Communities and Local Government (2013) *Helping Troubled Families Turn their Lives Around.* https://www.gov.uk/government/policies/helping-troubled-families- turn-their-lives-around.

Department for Education (2012) 'Government publishes destination data for the first time', 17 July. https://www.gov.uk/government/news/government-publishes-destination-data-for-the-first-time.

Department of Health (2013) *Reforming the Social Work Bursary: The Government Response to the Consultation*. https://www.gov.uk/government/uploads/system/uploads/attachment_data/file/203549/Reforming_SWB_-_Govt_response.pdf.

Dietz, T. and Kalof, L. (2009) *Introduction to Social Statistics*. Oxford: Wiley-Blackwell.

Dunstan, S. (Office for National Statistics) (2012) *General Lifestyle Survey Overview: A Report on the 2010 General Lifestyle Survey*. http://www.ons.gov.uk/ons/rel/ghs/general-lifestyle-survey/2010/general-lifestyle-survey-overview-report-2010.pdf.

Eurostat (2011) *Percentage of Women's Employment which is Part-time, in Four EU Countries, 2011*. http://epp.eurostat.ec.europa.eu/tgm/refreshTableAction.do?tab=table&plugin=1&pcode=tps00159&language=en.

Eurostat (2013) *Acquisition of Citizenship*. http://epp.eurostat.ec.europa.eu/tgm/table.do?tab=table&init=1&plugin=1&language=en&pcode=tps00024.

Faherty, V. (2008) *Compassionate Statistics: Applied Quantitative Analysis for Social Sciences*. Los Angeles: Sage.

Field, A. (2013) *Discovering Statistics Using IBM SPSS for Windows*, 4th edn. London: Sage.

Fielding J. and Gilbert, N. (2006) *Understanding Social Statistics*, 2nd edn. London: Sage.

Fisher, R. and Yates, F. (1974) *Statistical Tables for Biological, Agricultural and Medical Research*. Harlow: Longman.

Frankfort-Nachmias, C. and Leon-Guerrero, A. (2014) *Social Statistics for a Diverse Society*, 7th edn. London: Sage.

Fuller, E. (2013) *Smoking, Drinking and Drug Use among Young People in England in 2011*. A survey carried out for the Health and Social Care Information Centre by NatCen Social Research and the National Foundation for Educational Research. http://www.ic.nhs.uk/pubs/sdd11fullreport.

Garner, R. (2010) *The Joy of Stats: A Short Guide to Introductory Statistics for Social Scientists*, 2nd edn. Peterborough: Broadview Press.

Gliner, J., Morgan, G. and Leech, N. (2009) *Research Methods in Applied Settings – An Integrated Approach to Design and Analysis*, 2nd edn. London: Routledge.

Greasley, P. (2008) *Quantitative Data Analysis Using SPSS: An Introduction for Health and Social Science*. Maidenhead: Open University Press.

Halpern, D. and Coren, S. (1991). 'Handedness and life span', *New England Journal of Medicine*, 324, 998.

Harker, R. (2012) *Children in Care in England: Statistics*, Standard Note SN/SG/4470. http://www.parliament.uk/briefing-papers/sn04470.pdf.

Health Protection Agency (2012) *HIV in the United Kingdom: 2012 Report*. London: Health Protection Services.

HECSU (2012) *Graduate Employment Stable despite Weak Economy and Public Sector Cuts, Reveals Latest Figures from HECSU*. http://www.hecsu.ac.uk/assets/assets/documents/news/WDGD_Oct_2012_Press_release.pdf.

Horsfield, G. (Office for National Statistics) (2011) *Family Spending – A Report on the 2010 Living Costs and Food Survey*. London: Stationery Office.

Huck, W. and Sandler, H. (1979) *Rival Hypotheses: Alternative Interpretations of Data Based Conclusions*. New York: Harper & Row.

Humby, P. (Office for National Statistics) (2013) *An Analysis of Under 18 Conceptions and their Links to Measures of Deprivation, England and Wales, 2008–10*. http://www.ons.gov.uk/ons/dcp171766_299768.pdf.

Ipsos MORI (2010) *2010 Election Aggregate Analysis.*

Kinnear, P. and Gray, C. (2010) *SPSS 18 Made Simple.* Hove: Psychology Press.

Kulas, J. (2009) *SPSS Essentials.* San Francisco: Jossey-Bass.

Lader, D. (Office for National Statistics) (2009) *Opinions Survey Report No. 41. Contraception and Sexual Heath, 2008/09.* London: Stationery Office. http://www.ons.gov.uk/ons/rel/lifestyles/contraception-and-sexual-health/2008-09/2008-09.pdf.

Lader, D. and Steel, M. (Office for National Statistics) (2010) *Opinions Survey Report No. 42. Drinking: Adults' Behaviour and Knowledge in 2009.* London: Stationery Office. http://www.ons.gov.uk/ons/dcp19975_50795.xml.

Levin, J., Fox, J. and Forde, D. (2013) *Elementary Statistics in Social Research,* 12th edn. Upper Saddle River, NJ: Pearson.

Levitas, R. (2012) 'There may be 'trouble' ahead: what we know about those 120,000 'troubled' families', *Policy Response Series No. 3.* Poverty and Social Exclusion in the UK.

Local Government Group (2011) *Local Government Earnings Survey 2010/11.* http://www.lga.gov.uk/payandworkforce

Longhi, S. and Platt, L. (2008) *Pay Gaps Across Equalities Areas,* Research Report 9. Equalities and Human Rights Commission.

Marsh, C. and Elliott, J. (2008) *Exploring Data: An Introduction to Data Analysis for Social Scientists,* 2nd edn. Cambridge: Polity Press.

Ministry of Justice (2012) *Ministry of Justice Annual Report and Accounts 2011–12.* London: The Stationery Office.

Mosteller, F. and Wallace, D. (1964) *Inference and Disputed Authorship: The Federalist.* Reading, MA: Addison-Wesley.

Norris, G., Qureshi, F., Howitt, D. and Cramer, D. (2012) *Introduction to Statistics with SPSS for Social Science.* Harlow: Pearson.

Office for National Statistics (2011) 'Population estimates by marital status', Mid 2010 Statistical Bulletin. http://www.ons.gov.uk/ons/dcp171778_244768.pdf.

Office for National Statistics (2012) Statistical Bulletin *Marriages in England and Wales, 2010.*

Office for National Statistics (2013a) *2011 Census: Key and Quick Statistics for Local Authorities in England and Wales.*

Office for National Statistics (2013b) *Families with Dependent Children by Number of Dependent Children in the Household, UK.* Produced by Demographic Analysis Unit, Labour Force Survey.

Office for National Statistics (2013c) *Crime Statistics, Focus on: Violent Crime and Sexual Offences,* 2011/12. http://www.ons.gov.uk/ons/publications/re-reference-tables.html?edition=tcm%3A77–290621.

Office for National Statistics (2013d) Statistical Bulletin *Live Births in England and Wales by Characteristics of Mother 1, 2012.* http://www.ons.gov.uk/ons/dcp171778_330664.pdf.

Office for National Statistics (2013e) *An Overview of 40 Years of Data – General Lifestyle Survey Overview – A Report on the 2011 General Lifestyle Survey.* http://www.ons.gov.uk/ons/dcp171776_302655.pdf.

Pallant, J. (2013) *SPSS Survival Manual,* 5th edn. Maidenhead: Open University Press.

Park, A., Clery, E., Curtice, J., Phillips, M. and Utting, D. (eds) (2012) *British Social Attitudes: The 29th Report.* London: NatCen Social Research. http://www.bsa-29.natcen.ac.uk.

Park, A., Bryson, C., Clery, E., Curtice, J. and Phillips, M. (eds) (2013) *British Social Attitudes: The 30th Report.* London: NatCen Social Research. http://www.bsa-30.natcen.ac.uk.

Portes, J. (2012) 'Neighbours from hell'. Who is the prime minister talking about? *National Institute of Economic and Social Research Blog.* http://www.niesr.ac.uk/blog/neighbours-hell-who-prime-minister-talking-about.

Robinson, C., Nardone, A., Mercer, C. and Johnson, A. (2011) *Sexual Health.* Health and Social Care Information Centre. http://www.hscic.gov.uk/catalogue/PUB03023/heal-surv-eng-2010-resp-heal-ch6-sex.pdf.

Rowntree, D. (2003) *Statistics Without Tears: A Primer for Non-Mathematicians,* 2nd edn. Boston: Allyn & Bacon.

Rowntree, S. (1901) *Poverty: A Study of Town Life,* 3rd edn. London: Macmillan.

Rugg, G. (2007) *Using Statistics: A Gentle Introduction.* Maidenhead: Open University Press.

Salkind, N. (2013) *Statistics for People Who (Think They) Hate Statistics,* 3rd edn. Thousand Oaks, CA: Sage.

Sapsford, R. (2007) *Survey Research.* London: Sage.

Social Exclusion Task Force (2007) *Families at Risk: Background on Families with Multiple Disadvantages.* London: Cabinet Office.

Strang, J. (1991) 'Left-Handedness and life expectancy'. *New England Journal of Medicine,* 325:1041–1043

Taylor-Gooby, P. (2001) 'Risk, contingency and the Third Way: Evidence from BHPS and qualitative studies'. *Social Policy and Administration,* 35(2), 195–211.

Treiman, D. (2009) *Quantitative Data Analysis: Doing Social Research to Test Ideas.* San Francisco: Jossey-Bass.

Trussell, J. and Westoff, C. (1980) 'Contraceptive practice and trends in coital frequency', *Family Planning Perspectives,* 12(5), 246–249.

UNESCO Institute for Statistics (2013) http://www.uis.unesco.org/datacentre.

United Nations (2011) *Update for the MDG Database: Contraceptive Prevalence.* Department of Economic and Social Affairs, Population Division. http://www.un.org/esa/population/publications/2011-mdgdatabase/2011_Update_MDG_CP.xls.

Vauclair, C.-M., Abrams, D. and Bratt, C. (2010) *Measuring Attitudes to Age in Britain: Reliability and Validity of the Indicators,* DWP Working Paper No. 90. London: Department for Work and Pensions. https://www.gov.uk/government/uploads/system/uploads/attachment_data/file/214388/WP90.pdf.

Walker, J. and Almond, P. (2010) *Interpreting Statistical Findings: A Guide for Health Professionals and Students.* Maidenhead: Open University Press.

Wiles, R., Bardsley, N., and Powell, J. (2009) *Consultation on Research Needs in Research Methods in the UK Social Sciences.* Southampton: University of Southampton, ESRC National Centre for Research Methods.

Worcester, R. and Herve, J. (2010) *Was it the Sun (and the Times) wot (Nearly) Won it?* http://www.ipsos-mori.com/newsevents/ca/506/Was-it-the-Sun-and-the-Times-wot-nearly-won-it.aspx.

World Bank (2013) *World Development Indicators.* Washington, DC: World Bank. http://data.worldbank.org/.

World Health Organisation (2012) *World Health Statistics 2012.* Geneva: WHO. http://apps.who.int/iris/bitstream/10665/44844/1/9789241564441_eng.pdf?ua=1.

Wright, D. and London, K. (2009) *First (and Second) Steps in Statistics,* 2nd edn. London: Sage.

Yang, K. (2010) *Making Sense of Statistical Methods in Social Research.* Los Angeles: Sage.

YouGov (2013) *Voting Intention.* http://cdn.yougov.com/cumulus_uploads/document/1or1j1cocr/YG-Archives-Pol-ST-results-03–050212.pdf and http://cdn.yougov.com/cumulus_uploads/document/lxcfy3g2a1/YG-Archives-Pol-Sun-results-080212.pdf.

INDEX